Six Sigma and the Product Development Cycle

Six Sigma and the Product Development Cycle

Graham Wilson

ELSEVIER
BUTTERWORTH
HEINEMANN

AMSTERDAM • BOSTON • HEIDELBERG • LONDON • NEW YORK • OXFORD
PARIS • SAN DIEGO • SAN FRANCISCO • SINGAPORE • SYDNEY • TOKYO

Elsevier Butterworth-Heinemann
Linacre House, Jordan Hill, Oxford OX2 8DP
30 Corporate Drive, Burlington, MA 01803

First published 2005

British Library Cataloguing in Publication Data
A catalogue record for this book is available from the British Library

Library of Congress Cataloguing in Publication Data
A catalogue record for this book is available from the Library of Congress

ISBN   0 7506 6218 2

For information on all Elsevier Butterworth-Heinemann
publications visit our website at www.bh.com

Typeset by Charon Tec Pvt. Ltd, Chennai, India
www.charontec.com
Printed and bound in Great Britain by Biddles Ltd, Kings Lynn, Norfolk

Contents

About the author

Graham Wilson works behind the scenes with executives in a small number of well-known companies as they progressively transform their organization and, in turn, themselves.

He originally studied behavioural science and was awarded his PhD by the University of Bristol. After a spell at a leading London teaching hospital, he joined Exxon, where he worked as an internal consultant specializing in workplace change, employee empowerment and quality improvement.

His career has taken him around the world and continues to provide a wealth of exciting and unusual situations in which to learn. He has been a part of the leadership team of a number of start-ups and, in an interim role, within a number of established companies undergoing radical transformation. While he balances a portfolio career involving writing, inspirational speaking and coaching, he is a non-executive director and serves as a charity trustee.

Graham has a particular interest in the strategic opportunities that the future holds for organizations and individuals. He can be contacted via his website (www.grahamwilson.org).

Preface

Some management concepts seem to be so pivotal at the time they are first mooted and yet all too soon they disappear from the vocabulary and the next one takes over.

At first sight six sigma ought to be one of these. It is a tough concept to understand, being rooted in statistics. The standard it demands is hard to imagine happening in most organizations. The kind of mindset that is needed to make it work is dedicated and unswerving. It is not an overnight fix, typically taking several years before an organization can really claim to have 'made it'.

Twelve years ago, in 1992, I do not think many observers would have thought that six sigma would last more than five years or so. Nevertheless, it seemed to me at the time that this was an important approach. At the time I was working as an organization development specialist across several industry sectors. Despite six sigma having particular appeal to people of an engineering persuasion, the issue that my clients kept returning to was its potential to eliminate so much of the cost of developing new products. With one or two in particular, they had no alternative but to focus on this aspect, and soon we found ourselves breaking new ground. There were a few others trying to apply the six sigma approach to design and development, but we soon found that they were missing a trick. They had not thought of integrating more than one of the sophisticated tools at their fingertips together. What we did was to link a way of gathering detailed insights into customer needs, to a tool that would optimize the products or services to meet these needs at the lowest practical cost, to one that ensured that this performance was maintained.

We applied the approach in the nuclear industry, in motor manufacturing, in an assessment of the potential for transforming the inland mail, and in three different 'emergency response' organizations. It always needed adapting, and some relied more heavily on certain aspects than others, but fundamentally it worked.

In 1993, I wrote a book about the approach, called *On Route To Perfection*. I did not expect it to be an overnight best-seller. In the intervening years, I have written five others, and they all sold well, being translated into a dozen or more languages and produced in a couple of editions, but *ORTP* continued to quietly sell for a decade or so. I began to wonder why and did a little research.

Six sigma has continued to be a sound management approach. In particular, it has been very popular with multinational companies, and especially with those whose manufacturing bases are in the Asian and Pacific rim areas. My book, it seems, was doing very well in those countries particularly.

In the meantime I had gone through quite a transition myself. I had invested a lot in my own development, especially in the whole area of human development, and was primarily working as a coach to senior managers. Then, out of the blue, I was asked to work with a number of executives of a financial services company in Europe who were implementing a six sigma process.

My reservation with six sigma has always been around the plethora of 'experts' who were involved in some relatively restricted way in a project with one company, and who then try to apply the same ideas in a completely different organization. I soon discovered that this was very much the case for the company by which I had been approached. The simple tools and techniques were in place, but the executives had not bought in adequately to make it work.

With a revitalized interest in six sigma, and by now a lot more wisdom about the process of transformation in organizations, I thought it was time to revise the original approach, to bring it up to date, and to offer it in a way that may appeal to today's management teams.

So here is a substantially rewritten account of the integration of quality function deployment, Taguchi's methods of experimental design and statistical process control. I have not tried to write three textbooks in one, and you will find some of the approaches a little quirky: the important thing is that they work. One academic who reviewed my proposal felt that there needed to be more tools included and then suggested one or two; I am afraid he had missed the point: this is a book about six sigma and product (which includes service) development, not a comprehensive book of quality techniques, of which there are some excellent ones already. I have deliberately not included the basics of problem solving, which are essential in working towards six sigma, because I have already written a whole book on these (Wilson, 2000). Nor have I spent much time exploring the detail of the management of change because, again, I have written already on this (Wilson, 1995). I hope that you will add this to your repertoire of approaches, and that you will let me know of your successes: I love hearing from people and am happy to discuss any aspects of how you intend to apply, or are already applying, this approach to your work.

References

Wilson, G. (1995) *Making Change Happen*. London: FT Pitman.
Wilson, G. (2000) *Problem Solving*. London: Kogan Page.

Acknowledgements

In the first edition of this book, I included a separate chapter on Motorola as a case study. The material was very kindly reviewed by Shelagh Lester-Smith, Tonnes Funch and Bill Wiggenhorn of Motorola. As a result, I was invited to spend a considerable amount of time meeting and discussing matters of organization development with a wide range of fascinating people throughout the USA. I am indebted to them for their support and encouragement.

This book draws on my experiences with many clients. I thank them for their constant source of fascinating challenges.

On the subject of support and encouragement, my partner, Gilli Hanna, has had to put up with my prolonged immersion in this revision. Despite her understandable reservations, she has kept me motivated throughout. We both look forward to rediscovering the world outside the laptop together!

At Elsevier, Maggie Smith was quick to respond to the idea of this book, and her colleague Francesca Ford has been the one who has listened to my interminable excuses for not having it quite ready yet!

Thank you all.

1

Culture

I am going to begin with a highly personal perspective on the evolution of the six sigma movement. This is not going to be a soft sell for six sigma. If I put you off pursuing this process, then the book will have been well worth your investment in it. If I demonstrate that the process is much more complex than you had expected, then I will have grounds to add another star on my fuselage. Six sigma is a highly worthwhile goal. But it is not to be undertaken lightly: the effort involved will reap profound rewards, but the pain is too great for many organizations (or the leaders of them) to endure.

There has been a revolution taking place in business. The world of work has never before had the flavour that it has now. And it never will again.

A time in which minds were expanded, but products lagged behind

Since our minds would struggle otherwise, when we describe the history of business, we often talk in terms of decades. Some people are still at work today who worked in the 1970s. This was a decade characterized by design, innovation and industrial disputes. Ask most teenagers today when the Vietnam War ended and they will say it was in the 1950s, although actually it did so in 1975.

Innovation and design remain crucial elements in the mixture of success-making ingredients for business. Of course, the 1970s did not have a monopoly on them. (It is sometimes interesting hearing someone describing a stylish chair as so '70s' when it dates back to the 1930s. Innovation blossomed in both decades.)

The search for human beauty

Each decade seems to contribute something. While the 2000s might be seen as the era of technology, there has also been a dramatic growth in demand for antiquity, craftsmanship and the artisanal skills that were heralded by William Morris in the 1880s. Experiments in open-studio working today are replicating the model of the arts and crafts community that he set up in the Cotswolds.

The first step: appreciating quality

In the 1980s, in some ways people had grown tired of innovative products, manufactured quickly, using new materials and processes. Not because they did not like colour television, low-cost microwaves and budget hi-fi, but because they did not like them breaking down all the time. This was the world of the adults who conceived six sigma.

Behind the scenes, in the defence industry, and very much a consequence of the Vietnam War and other conflicts at the time, there had been a silent revolution brewing. The defence procurement specialists had evolved a code, a series of standards that could be applied across the board to ensure that suppliers delivered items that worked and were reliable.

In 1979, the first civilian equivalent of these military standards was launched. It was called BS5750 and it eventually evolved into ISO9000. It heralded a new era for the 1980s, of quality. This was not to say that the Quaker manufacturers of the 1870s, or the engineers who produced the first Fords, had not understood quality, but in the 1980s it became the buzzword. Suddenly everything to do with management had some connection to quality. To some extent this was a natural coalescence, but it was also a bandwagon.

Depending on who they were or what they did, individual managers clutched at a particular thread. (And boy, are some still clutching!) Engineers, outside the production area, seemed to resonate with quality management systems (as the ISO9000 fraternity called them). People in production loved the tools and techniques of quality control. (Its proper name, statistical process control, was clearly not hip enough.)

When human values matter

In the background were a few people who were pointing out that one-solution answers were unlikely to work and a more holistic approach (well, they would have called it that in the 2000s) was needed. One such guru of the time, Tom Peters, espoused four characteristics of 'excellent' companies: customer obsession, employee empowerment, transformed leadership and innovation. The formula still works, but it is, and always will be, 1980s' speak.

What Peters was saying very clearly (and he had a powerful style of oratory) was that excelling in business was as much about how you relate to your employees as it is about the nuts and bolts. His message, too, had its precedents. In the 1920s, Dale Carnegie and his associates penned *How to Win Friends and Influence People*, in which they outlined the basic principles by which people could work better together and by which managers could lead their staff. In the 1960s, especially in the USA, there had been the 'quality of work life' initiative that promoted engagement of the brains as well as the brawn of employees. (Had this not been sponsored by the organized labour movement it might have had more impact, and now that the nature of the relationship between unions and management is radically different, perhaps it is time for a forward-thinking union to try again.)

Understanding people at work

The first half of the twentieth century had represented an unfolding of the science of psychology. By the 1940s, a lot of attention was being paid to the interrelationships between people, especially at work. Psychologists were employed in the war machinery. Little did anyone know of the extent of this 'research'.

In the 1910s, early experiments by one man, Taylor, had led to the 'discovery' of scientific management or, as it is sometimes called,

'Taylorism'. The idea was elegant. Divide work into simple, short, repeatable units and have workers perform them for prolonged periods. In this way they become both proficient and efficient, and the overall process becomes highly economical and of the best quality. The trouble was that people did not like doing the same mindless task hundreds of times a day.

The result was riots: literally. Scientific management was outlawed by Act of Congress in the USA, and then under French law after further riots there. Despite this there are still some managers trying to implement it today.

Scientific management was a precursor to the science of organizational behaviour, but this had to go through one more desperate evolution before it would be applied appropriately. It was only in evidence at the Nuremburg war trials that the work of Wirth, especially in designing the processing plants at Belzec and other concentration camps, came to light. He had created a system, operated knowingly by human beings, with the sole purpose of killing and disposing of as many people as possible in as short a space of time as possible.

So we entered the 1950s. Psychologists knew that there must be positive applications of their work in peacetime. One man stands out above all the rest as the father of organizational development (OD): Ed Schein. Organizational development is perhaps best defined as the application of the science of organizational behaviour, but with a clear set of humanist values underlying it.

Essentially, organizational behaviour scientists were saying, 'It is all right to intervene in the way people behave at work, provided that they are enriched by the experience and not manipulated purely for the corporate advantage'.

Many critics of the 'change programmes' of the 1990s pointed out that they did not generally have the enrichment of the employees among their criteria for success. They were therefore manipulative and, in the worst cases, clearly corporate bullying in disguise.

If OD was the buzzword of the 1970s, in the 1980s it was called 'total quality management', and in the 1990s the 'management of change' (not to be confused with 'change management', which is an IT term). Since then it has diversified, and today we hear a lot about (inner) leadership, authenticity, emotional intelligence and spirituality at work. In each case, they are essentially about creating a workplace that is enriching for employees and at the same time successful for the organization. The themes mentioned are all reflections of the need for managers to have a more grounded sense of themselves and their own values before they attempt to manipulate people at work.

Listening to customers; anticipating their needs; building a relationship

If the 1980s saw a focus on quality, what next? It really did take the likes of Tom Peters shouting at management audiences for them to realize that quality means giving customers what they want. Even then, in the 1980s, many tried to avoid this by focusing on internal customers (to the exclusion of the ones outside who parted with cash). Most quality consultants of the time will tell you that the hardest group to influence through total quality management were salespeople: sales managers and sales directors. With hindsight, many were right to be sceptical, for their organizations focused exclusively on internal issues and paid lip-service, if that, to the real world of the buying customer.

By the 1990s, it was trendy to be totally focused on customer satisfaction. Sadly, a few companies were sold the idea that quality management systems would achieve this, and were rudely awakened when they discovered that staff who had boring jobs, and who were underpaid and bullied, rarely delivered satisfaction to their customers, regardless of the bureaucracy surrounding them. An industry was born servicing these needs: workshops, books, experiential groups and even, in one case, a company selling mirrors with a half moon on them, to which you were meant to match your grimace before speaking to punters.

As we entered the new millennium what began as an obsession, something with passion, had been systematized, regulated, legislated and turned into the field of 'customer relationship management'. In a number of cases, so important is this relationship with their customers that companies have transported their 'customer service' professionals (the jobs, if not the people) to the other side of the world on the basis that it is a lot cheaper there. Of course, it was only a matter of a year or so before they realized that if you ask people who are earning £1500 a year (if they are lucky) to handle loan applications for ten times that amount they might just resent it! And people who resent things can be tempted to redress the balance in their favour. Brains, it seems, are not necessarily a prerequisite for corporate strategists.

'And what', you might quite rightly ask, 'has this got to do with six sigma?' Two things immediately come to mind. First, some people will tell you that six sigma is a thing of the past. It was relevant in an emergent US economy in the 1980s, but not since and nowhere else. The answer is that it is neither time dependent nor determined by the climate (economic or otherwise); six sigma is the natural next step for a handful of organizations, for which the culture is ready.

By the early 1990s, and certainly by 1992, half a dozen companies were in active pursuit of six sigma: ABB, Motorola, TI, IBM, DEC and Kodak. As an aside, despite this broad range of companies being involved, Motorola subsequently claimed that six sigma was a trademark of theirs. They continue to make this assertion very broadly (and in all probability many of the people who make it believe it). Yet, many companies had already embarked on the process and Motorola's claim was restricted to the 'fields of electronics and telecommunications equipment manufacturing'.

Secondly, it tells us something about the external climate in which six sigma emerged. (I will not say was 'born' because it most definitely was not an invention of the 1980s, but stems instead from the pioneering work of a few souls in and around the 1950s (Deming, Shewhart, Ishikawa *et al.*).

Since then, GE, Allied Signal, Nokia, Sony, Navistar, Whirlpool, Bombardier, GenCorp, Siebe Foxboro, Lockheed Martin, John Deere, Siemens, Compaq, Seagate, PACCAR, Toshiba, Dupont and Dow Chemical have all launched six sigma processes. More recently still, in 2000, it seems there were active six sigma processes at Air Products, Honeywell, Johnson Controls, Maytag, Praxair, Ford, Zurich Financial Services and Johnson & Johnson.

The chief executives' embrace

The world around us moves on. Companies (usually one or two members of the management team) embrace ideas that appeal to them. Some will be a century ahead of the game. Others will be a bit behind. What was right for Motorola in the 1980s appealed to Zurich Financial Services (ZFS) in the 1990s, and may suit your company in the 2000s. Just as the approach adopted by ZFS was very different from that applied at Motorola, so the approach that you adopt will be different to either of them.

Organizational maturity

When a company embarks on six sigma and comes to a halt, it is not unusual for the problem to be that they tried to copy too closely the approach of one of the exemplars, they were not discerning enough or they were sold a mechanistic process developed by someone who only knew part of the story. Into this last category, I would especially lump the firms who will sell you 'black-belt' courses in problem solving. Not only are these short courses an insult to the martial arts fraternity, but they are also an insult to the intelligence of managers and staff in organizations. Fortunately, a few courses are emerging that do teach the practical application of human behavioural science and that treat their students as articulate, intelligent human beings. These are typically part-time MSc courses, lasting a year or more. (I mention these in more detail in Chapter 3.)

You will be glad to know that this book will describe a set of tools that have been used to good effect in companies in pursuit of six sigma, and show you how to integrate them, especially in the area of (new) product development. But it expects you to be discerning in your application of them, and not slavishly to follow someone else's (my) approach.

Six sigma, as you will see later, is a statistical term. The tools that I describe were all developed and shared among organizations seeking to achieve better levels of productivity, often from customer to supplier. There are some individuals, masquerading as organizations, who would seek to label this integrated approach and stick a copyright symbol on it. To my mind they are not only exceptionally immature and sadly lacking in more appropriate measures of their own self-worth, but also missing two important points: first, pursuing six sigma, or whatever you wish to call it, is not a competitive thing. It is about your own performance, if that performance gives you a competitive advantage that is a bonus. Secondly, it is out of sharing and collaboration that most innovation emerges. If you want to get ahead, get out there and share. If you want to die off through stagnation, then keep it all to yourself.

While I do not endorse national awards and similar schemes for organizations, the US national quality award, 'the Baldrige', has an important provision among its terms: winners are required to do what they can to promote their own experience among other organizations.

The culture of a six sigma company

Without doubt, there is no definitive culture, but there are some similarities.

The 'best possible' mindset

These companies do not embark on six sigma if they have a *laissez-faire* attitude. They know that doing anything more than once to get it right is bad economics. They know that a production line running at half speed (or even at 95 per cent of capacity) is not only wasteful but also has a direct impact on the bottom line. They know that an employee who comes to work but leaves their mind on some domestic crisis is not going to work their best. The leaders of these organizations work hard to create a mindset among their staff that they will do their 'best possible' work, and it is the responsibility of the managers to enable them to do so.

The driver of change

Such organizations are not generally obsessed with their customers. That is a shame, because if they did they would be more fun. You will not see successful, fun companies – Virgin, Southwest Air, Starbucks, even Disney – pursuing six sigma. The kind of people who resonate with six sigma are those who like a studied, serious approach. They do not really like risk or much spontaneity. That might sound like a downer, but it fits most companies. The exceptions are the few that stand out.

What, then, is the driver of change in these companies? Wrong question. It should be 'who is the driver of change?' In every case that I have come across, the successes and the less successful, there has been one person who has championed this process. Why? Because they (and usually it is an owner or a very secure CEO) realize that what needs to happen is a fundamental change in attitude at every level and in every area of their organization.

I do not mean to sound critical, but the culture of these organizations is generally quite patriarchal. Motorola, for instance, used to be justifiably proud of its 'founder's touch'. And when the chips are down, it calls for someone with such resilience to act.

Customers first?

Ten years ago, I would have jumped on the bandwagon and preached the gospel of customer satisfaction – and I did. What I have learned over time is that even these six sigma/excellent companies do not take customers that seriously. I am sorry, I know that is painful. I know a few people will recoil in horror. The sad reality is that despite all the good words about customers, many companies do survive, in fact excel, although they are totally customer insensitive. Contrary to the mythology, customer satisfaction is not a prerequisite for corporate success in this day and age. We are in an ever expanding marketplace. Loyalty of customers is nice if it happens for nothing, but most firms are focused on such extreme growth that they cannot afford to work on minimizing attrition due to dissatisfaction: they can only achieve the growth they want by short-term gains of large volumes of business.

A conviction in the potential of people

You may not agree with his methods (few do these days), but Jack Welch believed in the potential that people have to choose to do an exceptional job. It is little surprise, then, that he led six sigma through GE. Sadly, some people do not respond at first. Some have had a pretty tough time in life, and asking them to transform themselves is not likely to be very successful until they are ready to do so. They can, but they need time. However, Welch gave them a year and then said goodbye to the 'bottom' 10 per cent. By contrast, probably one of the most grounded, wise men of business who I have had the privilege to meet was a reformed drug user who drove a forklift truck in one of Motorola's plants. Part of his job went unsupervised and unaccounted for, as he was also the resident substance-abuse counsellor. If you ever need evidence of the power of people to turn themselves and their contribution around, he was it.

The one thing that six sigma companies do have, albeit in a patriarchal, sometimes even patronizing manner, is a conviction that the people who work for them can do better, and better, and better. They are the true investors in people, not the ones who go for a plastic shield on their office walls. They do it because it makes them feel better. They do it because they like to see people grow. They do it because people who are growing enjoy themselves, and people who enjoy themselves are generally better to be around. And they do it because they know that it leads to the long-term survival of their organization. They do not need evidence, they trust their intuition.

In the 1980s, Ralph Stayer transformed the nature of his sausage company, Johnsonville. The upshot was a plant run entirely by graduates. No one was fired. Few were recruited. Painstakingly, each person was helped to 'get' education. College professors taught in the works. Jobs were rotated to give practical experience. Pay was frozen; the only way staff earned more was if they learned more. Later, Ralph was interviewed. No, he explained, there was no 'road to Damascus conversion'. He had just begun to realize that people did not smile at work and that it was his job to do something about it.

Organizations that follow six sigma do not do so just to make things better. They do so because they believe in the potential of the people that work for them, they are prepared to invest in those people and they have the wisdom to follow through once they have started.

Think hard before you go further. If you are tempted to latch on to a few tools and techniques, if you feel that you are not empowered to transform the culture of your organization, then stop and think carefully. In Chapter 3, I spell out the crucial elements that are to be found in every successful six sigma implementation. The first one is to create a forum in which the most senior people in the organization discover what happens in very different workplaces. Only when they have really had their eyes opened to alternatives will they be able to support you in achieving your ambition for the company.

Timing

Without doubt, when Motorola embarked on their six sigma process, the time was right. The same could be said for every successful implementer. At the time that this book is being written, Motorola is in turmoil. After three generations of Galvin at the helm, the latest has resigned and the organization has been left behind in its performance for a decade. An outsider has taken over. Much of the former culture of Motorola has been destroyed in recent years. Today would not be the right time to launch such a process as six sigma.

The rest of this book

So what are you about to embark upon? This book tries to offer another insight into the incremental process that is behind six sigma. In the next chapter, I will define six sigma. Armed with that background, we look at the common steps that companies follow to create the culture that is so important in leading to six sigma. This will take the form of some generics, but also details of the journey at Motorola and some other approaches. The last three chapters describe three techniques, in sufficient detail for you to use them, which taken individually will almost certainly transform your organization's performance. Before these, in Chapter 4, we spend a few pages exploring how the three techniques can be integrated together. You do not have to integrate them, but nowhere else will you read of the ease with which these classic six sigma tools can be brought together. If you were after a definitive list of quality tools, with step-by-step instructions for all of them, then that is not the purpose of this book.

2

What is six sigma?

This chapter contains some very useful definitions and ideas that will help you to understand the concept of six sigma and how it is applied. Six sigma comes from statistics, so a certain amount of statistical information is included and this is important if you are going to get the most from later chapters; however, I have tried to keep this to the minimum.

Consistency, taste and variation

Since the 1960s, more and more people have been travelling farther away from home on holiday. While a minority have tried to export their homeland and all its comforts with them, many have been prepared to put up without the little luxuries, such as fish and chips, stout and toilet paper, in the search for the original ethnic atmosphere of their holiday destination.

Of course, for a long time people complained about poor sanitation, poor medical care, poor this, that and the other. But in most places the tables have turned and standards have equalized, certainly throughout Europe.

The souvenirs and presents that people bring back from their holidays have also matured over the years. There are still the mass-production injection-moulding factories manufacturing Italian figurines, Greek urns and Spanish dolls, although they have moved from Taiwan, where they would no longer dream of making such items when they could be making higher value electronics and automotive components, to eastern Europe. There will be small differences, but most of the output from any one of these lines will be the same. The extent of the variation is small, even if the overall specification is poor and the taste questionable!

The expansion in the 1970s of stores such as Habitat and Heals was due, in part, to the 'better quality' ethnic products: hand-made urns, wicker baskets, rugs, blankets, toys, candles and so on. Costing much more than items purchased abroad, their perceived quality was often due to the fact that they were different, that they did vary, but that their production standards were higher (they lasted longer). Notwithstanding the 'retro' boom that we are experiencing at the moment, the tastefulness of these items may still be questioned, although that is personal opinion.

Today, specialist retailers offer ethnic items, individually different, but to increasingly higher standards of production, based on sound designs and manufactured by specialist factories. Variation to allow individuality is encouraged, but within tight criteria for durability and, through better aesthetic design, ever increasing standards of taste.

Whether we are concerned with postoperative recovery in a hospital, or the manufacture of low-fat spread in a factory, 'quality' has usually been defined in terms of a specification and how consistently it is met. Our obsession with consistency is fascinating, and no doubt one day psychologists will throw some light on it. Even baby books tell mothers how their product should be developing; if it deviates from the specification, they rush to the health clinic seeking solace. But this is not enough. Quality, in the eye of the customer, is a complex range of choices. Some need to be consistent and others to vary. With the emerging idea of designer babies, even this area is changing.

A similar phenomenon is found in the hotel trade. Despite the enormous degree of variation between people, all hotels provide a core service: a bed. Most add a few other services, such as a room and meals, although there have been very successful experiments in not providing these. In Japan in the 1980s, for instance, you could stay in a cubicle with a bed but no room. This experiment did not work and few of these refuges for stranded executives remain. In Britain, however, the chains of 'motor-lodges' that evolved at the same time, providing a bed and a room, but with meals supplied only in an adjacent café, are perpetually full.

The argument that quality involves consistency led to a number of very consistent hotel chains. Today, most major chains have a budget product, such as the Marriott Courtyards and Holiday Inn Express groups. The rooms, furnishings and facilities are very similar from one hotel to another. The executive hopping from one city to another and one continent to another is assured that the bed will have the same consistency and the wallpaper will have the same stripes.

As the boom in servicing mobile executives declined in the mid to late 1990s, a different breed of hotel began to emerge. Those people with control over their budget were increasingly seen clutching the *Good Hotel Guide* as a way of finding something different, something unique. Often these smaller hotels would lack items that the larger chains considered 'essential', yet there was no evidence that their occupancy rates suffered. Today, though, their popularity has subsided, as corporate discounting through online agencies has favoured the mass chains again.

It is important for us to appreciate that variation occurs all around us. It is a perfectly natural thing; without it life would be very boring. Variation not only brings quality of life, as in the cases of ethnic souvenirs and hotels, but can also convey product quality.

Consistency, accuracy and precision

Two other features of quality are closely related to consistency, and it is useful to be aware of them. They are accuracy and precision. If you are paying a courier to deliver a parcel from Brussels to London, you expect the organization to have a specification that says when it will arrive. For an overnight service, the specification may be for delivery by 9 a.m. the next working day. If you are going to rely on them, they need to be accurate, i.e. they should not go outside that specification by delivering at 10 a.m., 11 a.m. or later still.

Until the early 1990s, couriers were judged on specification, accuracy and consistency. Then a fourth criterion began to be applied: precision. Working hours began to be blurred, more people were self-employed and worked longer hours, many businesses no longer held to a 9 to 5 regime but expected their staff to arrive earlier and leave later, and globalization meant shifts that crossed one another and interacted around the world. Couriers began to discover that 'before 9 a.m.' was no longer adequate as a specification. They had to be more precise, 'between 8.30 a.m. and 9 a.m.' To its cost, one courier simply tried to shift the specification, but then began to lose business because it was trying to deliver to offices in London at 7.30 a.m.

The UK postal service struggled to keep up with this moving feast of customer expectations. Its specification was a pretty unimpressive, 'by 12 noon', but it survived by charging a fairly low cost to achieve this. It was, to all intents and purposes, pretty consistent in achieving this, but it began to lose business as the importance of earlier delivery outweighed the cost advantage. Since its relaunch as the Royal Mail, it has had to offer a two-tiered service to meet the needs of its customers. It effectively offers a product that has two specifications. However, it still has to resort to extraordinary means in some places to meet the tougher of these. Being precise and accurate is essential, but the service must be consistently so. It is no good if nine times out of ten they deliver by 9 a.m., but then foul up one in ten times. If we knew which delivery they were going to foul up, perhaps we could live with it, but this will not be the case. It could be the next delivery, or the next, or the next, and so on.

The old analogy of two Western gunfighters is useful. In the 1880s two gunfighters fought it out in a street. One was armed with a precision revolver, although he hadn't had it all that long and so lacked practice in using it. The other had an old and trusted weapon that scattered bullets, but he had used it many times before.

The first to draw had the new revolver. It spat out six bullets in quick succession. Had his arm been stationary they would have all fallen within inches of one another. As it was, his arm was still moving and they fell in a graceful and precise arc towards his enemy. Sadly, the chambers were empty before the precise arc could coincide with human flesh.

The second man took a more leisurely approach. Drawing his weapon, he pointed it in the direction of his victim and also let off all six chambers. His bullets went all over the place, with no precision at all, but because he was accustomed to handling the gun the centre of the cloud of bullets was pretty close to the victim's heart and one of the six managed to strike him dead. The gun was not precise, but it achieved its target and was therefore accurate.

Activity

It is worth spending some time mulling over what 'precision', 'accuracy', 'consistency', 'taste' and 'specification' mean in your organization. The differences can often be surprising. Of course, these are not mutually exclusive. For example, the specification can include measures of accuracy, consistency, and so on. If you want to remember them, try the acronym PACTS: what you make with customers!

Sigma is the statistical parameter used to measure variation, but before you apply it, you need to decide very carefully which of the five variations you are trying to measure, and what is the objective of that measurement. In Chapter 5, we shall look at how you decide which is important, or rather how to get the customer to decide for you. In Chapter 6, we show how you can improve and control these measures. In Chapter 7, we look at how you can sustain them and even save effort in doing so.

Variables

Imagine that you run a high-street travel agency. Your customers can call you by telephone or they can pop in off the street. Your staff ask questions, provide information and hopefully convert casual enquiries into firm bookings. A booking does not stop with the holiday flight and accommodation; there is also the chance to sell insurance and foreign currency, as well as car, bike or equipment hire and tuition. From the customer's perspective, foul-ups in delivering their holiday and its mix of products will probably result from mistakes on order forms, the loss or lateness of documents, and errors transcribed onto them, because that is what they associate with you.

For most of your work, however, you are acting as a convenient intermediary between the customers and various other service providers. Because of your proximity to the customer and the amount of information that you share, there is actually less chance of a foul-up on your part than on that of one of the main service providers.

You have developed a set of procedures, maintained by computer, to ensure that important information is properly documented, and that any problems brewing are easily spotted. These forms contain such information as the date on which a booking form was sent and the date on which the confirmation was received. They indicate when the travel documents were received and have tags to show that your own staff have checked the contents for accuracy.

Now you want to improve, and have set yourself a target for the end of the year. The target is to have a clear measure of how many transactions of each type you are processing and how many 'errors' occur in each.

Once the customer has booked a trip, you begin to gather their documents so that they can all be issued together. As they are received from the suppliers they are crossed off a list of outstanding items. Your definition of an 'error' is where a document is not received within, say, ten working days.

Whether it is managed by computer or manually, you have a visible list of outstanding items and their deadline date. As soon as something goes beyond its deadline it becomes an error. You could record this as an error in your management report and count the total number of errors in a particular transaction. This is an 'attribute' variable. Alternatively, you could record the number of days for which the items have been outstanding. This is a 'continuous' variable.

For some reason, many people think that statistics can only handle continuous variables. In fact, it can deal with both attribute and continuous variables, and in Chapters 6 and 7 we shall see how.

Continuous variables

Continuous variables are measured on scales. They are a type of scalar variable. Not all scalar variables are treated as continuous. For example, the energy emission of a hot body may be classified in terms of the colour of light generated. The scale goes: red–orange–yellow–green–blue–indigo–violet. The frequency of the light is a continuous scale, but it is much easier to refer to (and measure) colours.

Attributes

Attributes are often quicker and easier to collect and can often be handled more easily as statistics than continuous variables. An attribute is defined as any type of data that only has two possible values. Examples include: conforming/non-conforming, pass/fail, go/no-go and present/absent. This type of data can be counted quite easily, provided that the distinction between the two states is clearly defined. Because of their ease of use, attribute data are often collected routinely, even when it would be possible to make more detailed measurements.

For example, contraceptive sheaths are tested for the presence or absence of electrical continuity between the inside and the outside, although it would be possible to record the actual resistance, a continuous variable. Instead, the number of defective sheaths in a batch is recorded based on this pass/fail attribute. This is subject to quality control (see Chapter 7) and only if the attribute data show problems is more detailed information gathered. This saves an enormous amount of effort.

It is worth bearing in mind that most continuous variables end up being reported as attributes, or at least as discrete chunks along a scalar axis. For example, you may measure the temperature of a room and report it to the nearest degree. The scale of temperature is pretty well known, from 0 to 100 degrees, or whatever. Although you could record temperature to a fraction of a degree, it is not worth doing so, and therefore you measure it more roughly in one degree intervals. The scale remains the same, the measure is still continuous, but the precision is less. Suppose that you need to produce a graph of the data. For the purpose of plotting on a graph, you adopt a simple scale in five or ten degree intervals. Eventually you could end up with a scale of just two intervals, pass and fail: the continuous variable has become an attribute.

Activity

For your own part of your organization, what kinds of data are collected? Are they attributes or continuous variables? What other kinds of information could be collected if you felt the need?

Distributions

One of the easiest charts to produce and one that can be really useful is known as a 'stem and leaf' diagram. Figure 2.1 shows the data from a motoring association. It records the time taken for the first patrol to arrive at the scene of a breakdown on a stretch of the M25.

The table of data has been transposed onto a horizontal barchart in the form of tally marks. This gives a very useful visual impression of the data and how they vary. As a rule of thumb, it is best to organize the data so

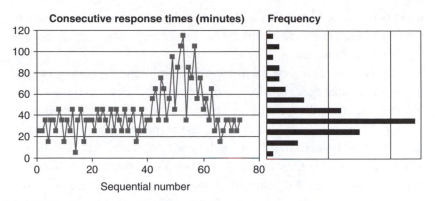

Figure 2.1 Response times (M25, junctions 6–11)

that there are ten equal classes. If we had done this they would each have been twelve minutes long, which would not have been too bad as there would then be five classes per hour; however, we have used ten minute intervals as they are easier for most people to relate to. For most purposes it is better to use a sensible division rather than sticking to the rule.

In this instance we can see that there is a range from less than ten minutes to almost two hours before the patrol reached the stranded motorist. The chart shows that the most common time for anyone to wait is between thirty and forty minutes. This is called the mode. The distribution of the times is said to be unimodal because there is only one such peak. (In fact, there is a slight glitch between 100 minutes and 110 minutes, but it is not really big enough to worry about here.)

The tail-off from the mode is steeper on the quicker side than on the slower side, and a statistician would describe this data as skewed negatively. (The left-hand side of the mode is usually considered to be negative.)

Most commonly used statistical tests assume that the distribution of all data is of one particular shape. This shape is unimodal and not generally skewed. It has a bell-like appearance and is called the normal distribution. If the leaf plot that you produced clearly had more than one mode, then you would think twice before carrying out the statistical procedures described later. (This is, however, very unusual, for reasons that will be explained.)

If a distribution is perfectly symmetrical, then the average (more properly the arithmetic mean) will be the same as the mode. In our case, because of the negative skew, it will be slightly higher (more positive). If the data refer to rates (such as flow rates or speeds), then the arithmetic mean is not quite right either and you should refer to a statistics book to see how to use geometric means. Again, in the real world of business this would be very unusual.

In the case of a symmetrical distribution, arranging the data in ascending or descending order results in the same number of data records to the left as there are to the right of the mean. So the value at which 50% are more and 50% are less is also the same as the mean. This value is known as the median. Looking at the data in the chart, there are 75 calls recorded. The middle value is therefore call number 35, as there are 37 below it and 37 above it. Counting upwards from the lowest, the 35th data record is in the class 0:30 to 0:40. In other words, although the data are skewed, the median is very close to the mode. Certainly it is in the same class.

Why bother with medians? Well, the median is much less affected by freak values than the mean. For example, the ownership of the shares of a company often fits what is known as Pareto's principle. Eighty per cent of them are owned by 20% of the investors. Thus, of 100 shares split among five people, one person might own 80, while the other four individuals may own only five shares each. The mean shareholding is 20, but arranged in ascending order (5, 5, 5, 5, 80) the median shareholding is 5. If we are concerned with serving the 'typical' shareholder, the person holding five shares would be a much better model.

Unfortunately, if we are taking samples of data from a population, the median will tend to vary more between samples than the mean does. Because most statistical work involves samples, the mean is used much more often than the median.

The mean, median and mode are all statistics to describe the central tendency of the data.

Dispersion

We have already used two terms to describe the dispersion of the data, when we said what its range was and that it was skewed. The mathematical formulae that can be used to describe skewness will not be described here. Instead, we usually use some measure of the spread of the data. The range is the easiest. If a quantity of data comes from a perfect normal distribution, then we can just state how many values there are, the average and the range, and a statistician would be able to re-create the numbers.

Unfortunately, though, data rarely come from a perfect normal distribution, so a better measure of dispersion than the range is used.

The next easiest method is to take the individual data values and subtract them from the average, ignoring the plus and minus signs. (If you do not ignore them the total will be nothing because that is how averages are defined.) Table 2.1 shows this calculation: it is called the sum of the deviations. In a collection of numbers that is widely spread, this number will be large, whereas in one that is narrowly spread it will be small.

The exact value of the sum of the deviations will depend on the scale of measurement rather than on the shape of the distribution. So, if you are measuring a set of data on a scale of 0 to 1, the sum of the deviations will be small compared with that on a scale of 0 to 100. This makes comparisons of distributions very difficult, so statisticians have a different

Table 2.1 Measures of dispersion

Compare two samples from the same population	
3, 4 and 5 with	1, 4 and 7

The average is the same: 4

The range is different:

$5 - 3 = 2$	$7 - 1 = 6$

The sum of the deviations from the average is:

(with signs)

$3 - 4 = -1$	$1 - 4 = -3$
$4 - 4 = 0$	$4 - 4 = 0$
$5 - 4 = 1$	$7 - 4 = 3$
Total $= 0$	Total $= 0$

(without signs)

$3 - 4 = 1$	$1 - 4 = 3$
$4 - 4 = 0$	$4 - 4 = 0$
$5 - 4 = 1$	$7 - 4 = 3$
Total $= 2$	Total $= 6$

The sum of the squared deviations is:

$(1^2) + (0^2) + (1^2) = 2$	$(3^2) + (0^2) + (3^2) = 18$

The variance is:

$[(1^2) + (0^2) + (1^2)]/3 = 0.66$	$[(3^2) + (0^2) + (3^2)]/3 = 6$

The standard deviation is:

$\sqrt{(0.66)} = 0.81$	$\sqrt{(6)} = 2.45$

figure that they use. Instead of just taking the sum of the deviations, they 'standardize' it. This is done by taking each deviation from the average and squaring the difference before adding them up. This removes the plus and minus signs (because a minus times a minus is a plus), and also biases the answer in favour of the larger numbers. For example, if the average was 4 and one individual data record was 2, the difference would be 2 and this squared would be 4. With the same average a record that was 3 would have a difference of only 1, which when squared is also 1. So this standardized deviation will be larger when the numbers are more widely spread, and is therefore a much better measure of how spread they are.

If you have a sample of ten measures from one place and 100 from another, the second set will produce a larger total. So, not surprisingly, we divide the total by the number of deviations we have calculated. This is known as the variance of the sample and is represented by the symbol s^2. The variance of a population is represented almost universally by the Greek symbol σ^2. Statisticians often use lots of subscripts too, but these symbols will do to understand any realistic textbook.

If only it were that simple! There is one further consideration. Because we squared the figures, the scale of our measure of deviation is different from the one on which the original was recorded. We must therefore take the square root of the variance, and this is the standard deviation. This is represented by the symbol s (or SD) for a sample, and the Greek letter sigma, σ, for the population as a whole.

Fortunately, slide rules and logarithm tables have all but vanished from our offices. Even ten years ago, we would have needed to use a scientific calculator instead. Five years ago, there was a range of spreadsheet packages, but today there is realistically only one. Once we know what the symbols and terms mean we can forget about having to calculate them and rely on Microsoft Excel instead. Each release of this package incorporates even more complex functions to the point that today's spreadsheet user has access to more computing power than a mainframe computer twenty years ago. Of course, having access does not mean having to use it. For six sigma work, you will probably only use half a dozen functions. To begin with, stick to average, STDEV and VAR.

Most of the time we are concerned with looking at samples rather than the whole population. This is true whether we are concerned with the number of typographic errors in correspondence, the time it takes to respond to an emergency call or the lengths of precut timber being delivered to a building site. If we were to measure each one we would simply add to the boredom factor and increase the chances of human error, and all that inspection would be a cost of quality that we could also do without. Interestingly, with the growth in the use of computers for automatic data collection, in a growing number of situations the 'sample' is actually every occurrence. There is probably next to no benefit from all this data recording, but since the systems are there we use them. Often it takes more effort to sample it than to treat the whole lot.

We have already seen the symbols used when we report the standard deviation or variance of a sample and of a population. Technically, we can talk of the sample as being an estimate of the population. Different symbols are used for the mean of a sample (x or $\bar{x}$) and of a population (μ).

If we had the measurement for a particular variable for every member of a population, then we could calculate that population's mean and standard deviation. If we remembered these two figures, but lost the original data then, in theory, if there were 100 figures we could dream up 99 and only one more would have to be calculated to find the same value of mean and standard deviation. This seems a rather unlikely situation, but it is the sort of thing that keeps some people occupied for years!

In other words, for a population of n individuals there are $(n - 1)$ opportunities to change the data without affecting the end result, since you can fix the total with the last item. Statisticians call these opportunities 'degrees of freedom'. If we are calculating the standard deviation of a sample using the formula described above, then instead of dividing by n, we divide by $(n - 1)$ to obtain a better estimate of σ. The larger the size of the sample, n, the smaller the difference produced by dividing between n and $n - 1$. For samples of over 30 the difference is meaningless. So, why bother with all this?

Fortunately, most of us these days do not bother checking the calculation performed by a spreadsheet on a set of data. Microsoft, however, includes different functions for the different standard deviations and variances in the Excel program. If you did not know the difference, it would be easy to use the wrong one unwittingly and end up with egg on your face!

The functions in Excel are:

■ STDEV: used to calculate the standard deviation of a sample where the array in a spreadsheet may have text values in it

- STDEVA: used to calculate the standard deviation of a sample where the array does not have text in it
- STDEVP: used to calculate the standard deviation where all members of the population have been measured.

In the extremely unlikely circumstance that you are using a calculator, it will probably have two buttons, one marked 'n' and the other '$n-1$', for both the mean and the standard deviation. Use the one marked '$n-1$' and you cannot really go wrong.

If you think that this is nit-picking, hold judgement until you read Chapter 7.

Central limit theorem and the normal distribution

This is probably the most useful theory in statistics because it helps us to apply statistical tools in the real world. Earlier we described the symmetrical, bell-shaped distribution known as the 'normal distribution'. This distribution has some very specific features that make it particularly useful in industry and elsewhere.

The normal distribution was described independently by three scientists, De Moivre, Laplace and Gauss, in the 1730s. Even today there are a few people who prefer to call it the 'Gaussian distribution'. The property of the normal distribution that is of particular use here is that the proportion of data records falling under different points of the curve is consistent.

The distribution is based on the idea that there are more data records closer to the mean than at the extremes of the range. For instance, if the distribution was concerned with IQ, then most people, i.e. most data records, would be close to the average IQ of 100. The number of increasingly clever people would be decreasing, so there would be very few geniuses at the top end of the scale. Similarly, the number of people in the lower half of the scale would be decreasing away from the mean. Thus, there would be very few complete dunces. Of course, this assumes that dunce and genius are at opposite ends of the scale of IQ. Not many people would agree with this these days, so just take this as an illustration.

The mathematics of the normal distribution are quite specific. Within one standard deviation on either side of the mean there will fall 68.26% of the data records. Within two standard deviations on either side are

95.44%, within three on either side 99.73% and four contain 99.994%. Armed with the basic parameters of the distribution – the mean and standard deviation – we can predict how many records will fall between any two actual levels. For example, if we were to equip an army with desert clothing and knew the mean size of the personnel and the standard deviation, we could predict how many of each size we would need in our stores to respond quickly without having to measure all the people individually and without making them squeeze into uniforms that were the wrong size.

The normal distribution is the most frequently encountered distribution in nature. Its real beauty lies in the central limit theorem. This states that if samples of a known size are drawn at random from a population, then regardless of the shape of the parent distribution the sample means will tend to follow a normal distribution. Further, the means of the parent and samples will tend to be the same, while the standard deviation of the samples will be approximately that of the population divided by the square root of the sample size.

This is the mathematical relationship:

$$\text{Population SD} = \text{Sample SD}/\sqrt{(\text{Sample size})}$$

The larger the sample size, the closer the fit will be, although for practical purposes there is not usually a large increase in accuracy above 15.

Let us take a practical example. Table 2.2 shows the data from a factory producing cardboard disks for a specialist, safety-critical application. Disks are sampled in units of five from each batch. The table shows three consecutive samples of five data records each. The average of each of these samples is also shown.

So far, we have seen that samples taken from a population, which we could not possibly measure completely, can be used to estimate the shape of the population. Using these estimates, we can go on to make predictions about the proportions of the population that fall between any two points on the scale of measurement that we are using.

Six sigma

In the same way that the points that represent four standard deviations on either side of the mean encompass 99.994% of the data records, six

Table 2.2 Average and standard deviation (SD) for a three-batch sample

	Sample		
	A	B	C
	0.65	0.75	0.75
	0.70	0.85	0.80
	0.65	0.75	0.80
	0.65	0.85	0.70
	0.85	0.65	0.75
Average	0.70	0.77	0.76
SD	0.0866	0.0837	0.0418
Sample size	5		
SD of the averages	0.0379		
Actual SD of all 15 items	0.0753		
Estimated SD of the population	$0.0379 \times \sqrt{(5)} = 0.084747$		

standard deviations on either side contain 99.99966% of the data. In other words, only 3.4 data records in one million will be outside those two levels.

In Chapter 7 we look at how six sigma affects one other statistical parameter, 'capability'.

Probability

Suppose then that we have been regularly sampling the output from a production process. We have produced a leafplot that confirms that the samples are producing a normal distribution. We have an estimate of the population mean and standard deviation. Then one day we take a sample with a mean that is more than three standard deviations away from the estimated population mean.

Are we surprised? We certainly are! With 99.73% of the data falling between the three standard deviation limits there will be less than three in 100 records that do not do so. This is not necessarily enough to provoke a panic. If the next consecutive record is similarly extreme, then we have real cause for concern. In Chapter 7 we shall look at more examples of such probabilities.

3

The transformation that is six sigma

Undoubtedly, the paragon of the six sigma process is Motorola. Indeed, the company has even implied that they 'invented' it. In the first edition of this book, I included a separate chapter on Motorola as a case study. The material was very kindly reviewed by Shelagh Lester-Smith, Tonnes Funch and Bill Wiggenhorn of Motorola.

Although many organizations have embarked on a six sigma process since then, none has yet progressed as far as Motorola, so it still seems appropriate to continue to use them as a case study. However, at the time of writing (May 2004) Motorola is in considerable flux. Back in 1995, the third Galvin, Chris, took the helm of the organization. His father had maintained the 'founder's touch' through a long period of considerable change in the communications sector.

While Motorola invested strongly in securing its markets, in extending globally and in communications infrastructure development, its six sigma process was timely in that it addressed a significant quality issue. However, the focus had perhaps, with hindsight, been placed too heavily on product quality and on infrastructure and too little on consumer products. Some people are still using Motorola Micro-tac analogue handsets that must be a decade or more old. But the cellular phone market, on which Motorola had come to depend heavily, was to change.

Three companies in particular transformed it: Sony, Ericsson and Nokia. Recognizing the very low manufacturing cost of handsets and the value of tied-in customers to airtime providers, they effectively turned mobile phones from communications devices into fashion accessories. They then turned fashion accessories into fashionable, and indispensable, personal entertainment tools. Even those who prefer not to acknowledge their entertainment value are happy to accept them as gadgets!

Bob Galvin's wise leadership had not created a manoeuvrable ship and Motorola's fortunes slumped. In strategic terms, this is very common. Exemplars of strategy, it seems, often fall foul of short-term dramatic market changes, usually brought about by perceptual change rather than anything tangible. As a futurist, I often find myself working with organizations exploring the options open to them, yet they are often unable to envisage what we call 'wildcards' as having a profound effect on their strategy. And yet, that is exactly what landed at Motorola.

So Chris Galvin inherited a monument of twentieth century manufacturing, and yet one trying to operate in a marketplace that was dictated by fashion, rather than technological solidity. When firms are organized on an ethos of quality as Motorola most certainly was, there are some things that they will find it hard to do, and one of these is to respond quickly to fashion-led obsolescence. Sadly, many of Motorola's products, while technically sound and built to last a lifetime, were fashionably obsolete.

As their competitors vied with one another for rapidity to market, marketing and advertising profile, a clear leader emerged in Nokia. Even the merger of Sony and Ericsson only created a second in line. Motorola was clearly deposed.

This situation was not without precedent. Back in the late 1970s and early 1980s a war had been waged between the photographic film manufacturers. Very well-established market leaders almost vanished into obscurity, usually clinging on to specialist niches (Agfa-Gaevert and Ilford) or becoming the stable but unexciting sliced white variety of supermarket own labels (3M). Despite leading this market revolution, Kodak was caught unawares when, at the Olympic Games opening ceremony in Atlanta, the film of choice and primary sponsor was Fuji. This market has transformed again with the advent of digital cameras, and Kodak still appears to struggle to compete with Fuji, which currently (and cleverly) straddles the fence between the media providers and consumer camera market (where Kodak lies) and the technical and professional camera market (with the likes of Canon, Nikon and Olympus).

So Chris Galvin began a further transformation. Not only did he have to change the internal culture from solidity and enduring quality to a more responsive, fashion-led and disposable one, he also had to grow the business to new heights of sheer turnover (otherwise Motorola would have been carved up and sold off), but, at the same time, to reduce costs drastically. Market analysts had mixed views on how successful he was. It seems he turned around Motorola's losses but growth was not sufficiently fast for the board. In 2003, they announced that Chris Galvin had indicated his intention to resign once a suitable replacement had been found. By December, Ed Zander, formerly head of Sun Microsystems, had been appointed.

Zander's approach was interesting and highlights the importance of different styles of leadership for different situations. While Galvin's approach was direct, clear, painful and relatively uncompromising, Zander has 'disappointed' some analysts by refusing to say what he thinks Motorola should do, but instead insisting on listening to the people inside the organization. The parallels with the situation twenty years ago that led them into six sigma are fascinating.

So what was the beginning? Over the next few pages I have related the transition at Motorola with the stages that are described in an arbitrary empirical model of change that is explained in detail elsewhere (Wilson, 1993). This is represented by Figure 3.1. This model emerged from an analysis of the change processes in a large number of organizations. I do not pretend that this is my 'invention', nor am I so ego driven as to slap a © symbol on it, and for every company that applies this sequence, there will be more who do not. It is a convenient short-hand for a dynamic process.

A climate of open-mindedness

One of the first steps for most organizations that are about to embark on a culture change process is for the executives to spend time reviewing strategies and exploring the alternatives. To do so, they individually need to be aware of developments outside their industry. One common complaint about executives is the very limited time they allow for this crucial aspect of their own development. It is easy to assess in conversation with them.

At Motorola there was already a climate of open-mindedness about the future that had been cultivated by the founder, Paul Galvin, and nurtured

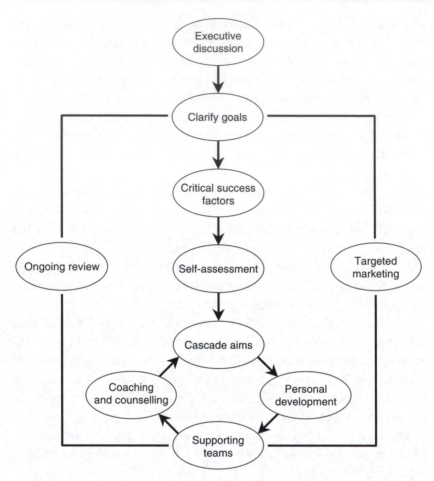

Figure 3.1 An empirical model of the organizational development intervention process

by his son, Bob. Since it was founded in 1928, an almost philosophical approach had evolved, which led the decision making and flavoured the business in many different ways. For example, few other organizations publish their own library of corporate and personal philosophical texts for distribution to employees. It might be argued that this is a vestige of the late 1920s, the time at which the Dale Carnegie organization was also founded; indeed, there are some striking similarities between the publications of Carnegie and Motorola. Whether this is the case or not, the culture within Motorola is one of discussion and debate. Ideas are never stupid; they are always worthy of exploration.

Organizations embarking on six sigma, or any other significant change for that matter, need to create this atmosphere first. Where I have been involved in, or led, such processes we have created an extensive programme of awareness raising for the executive team. We have included attendance at seminars and conferences, circulated interesting literature (books, videos, tapes and the like), organized exchange visits to other companies, encouraged participation in industrial professional bodies and a host of other activities. We always aim for members of the team to take part in pairs and try to debrief them thoroughly through coaching to achieve the maximum benefit. The context of the interventions is less important than the learning acquired from them, which is aimed at reinforcing the questioning approach where the status quo can be challenged.

Although this step may seem dispensable it is a vital one, and in some cases extends for many months. Without it, individuals can feel steamrollered and react negatively to proposals for change. In the case of one major UK manufacturer, a relatively new chief executive tried to implement a culture change process leading towards six sigma. Failing to invest in this step led to a complete loss of faith in the chief executive and he resigned shortly afterwards.

Within this climate, there will be many stimuli to change, and it may take only one further provocation to begin. This trigger can be provided by a customer or by someone internally. Sometimes it leads from research commissioned by the company.

At Motorola the obsession with quality has been widely ascribed to one person and his intervention at one management meeting. But that intervention itself said nothing particularly new; it was the climate of open-mindedness that allowed it to be made and to grow roots.

When Bob Galvin inherited Motorola in 1956, he had already had many years' experience. His father, Paul, was the entrepreneur: loud, impulsive, at times dogmatic and certainly someone with conviction. His son's approach was more analytical, cooler, more down to earth. Bob thrived on new technologies. Despite a tremendous growth in global competition, he turned a business of $227 million turnover in 1956 into one of $13 billion in 1992.

The event that triggered the six sigma process came during a company-wide officers' meeting. A senior sales officer named Art Sundry stood up. In almost any other organization what he said would have rung the death knell for his career. He said: 'This is fine. These are good topics and we're making some progress. But we're missing the point … our quality stinks!' The chairman at the time, George Fisher, later recalled that the group went through denial, trying to prove that he was wrong. But that event proved pivotal for Motorola.

After the initial remonstration by Sundry in 1981, Galvin and Bill Weiss (then CEO) launched the company on the first of many quality improvement goals: a ten-fold improvement within five years. At the time it seemed impossible, but it turned out to be insufficient.

Activity: Ongoing discussions

For your own organization, what kinds of executive development opportunities are offered? What is the take-up?

Document your own personal development over the past two years. Where has your knowledge been grown? Which new skills have you achieved and how? How have your attitudes changed and what has led to this?

Clarifying the goals

The next stage in the process is to reach a common understanding of what the new culture means and how it relates to the organization. The senior management team needs to reach consensus on their vision for the business, and identify some of the consequences of not working in this way. They will agree where they are now and where they want to be. Most importantly, they will become aware that new behaviour has to be adopted if the culture is to change, and that interpersonal skills will be needed to support this behavioural change. The time that it takes to arrive at this new vision varies, but eventually it can be encapsulated in a few short sentences.

The biography of Paul Galvin, published by Motorola, begins with the story of a twenty-year-old line operator in Malaysia who when asked what she liked about the company responded, 'The open door policy that was started by Paul Galvin' (Petrakis, 1991). She was born after Galvin had died, and yet the vision stayed on with remarkable clarity.

To take the organization towards its quality vision, Bob Galvin and his team needed not only to understand the issue of quality far better, but also to be convinced of its potential benefits. The quality movement was well underway, particularly in the USA, but Galvin's team spent longer in this phase of change than many others. There is a tendency for business people to be driven to do something. Nowhere is this more apparent than in the software and high-technology industries. Yet the team wisely resisted this temptation. While distilling the vision and their goals for the organization, they spent a lot of time considering the implications of their change process on senior and middle management.

While the organization beavered away at its first wave of improvements the top team was watching, listening and learning. Galvin recognized the value of training and made major commitments to ensure its provision. The top team was not exempt. They too had to learn and relearn. At their operating and policy committee meetings the quality process became a core agenda item. Throughout this time, for the top team, the goals of the process were only really being formulated.

Activity

Without referring to anything, describe your organization's vision and values.

Check this by asking two or three colleagues from other parts of the organization for their version.

Now have a go at formulating a statement of vision for yourself and add the values with which you would personally like your life to be consistent.

Finally, repeat this process with your partner.

Critical success factors

Having put into words their vision of the company of the future, the executive team will need to take more time to develop clear measures of success to monitor the process of transformation. These measures will involve several strategic decisions. It is often only later that the decisions may involve committing to the goal of six sigma.

In recent years, the model used to steer these definitions has often been that of the balanced scorecard, originally proposed by Kaplan and Norton (1992). In practice, most companies latch on to the four areas of interest rather than the meat of the authors' proposals. Kaplan and Norton recommended that firms should no longer focus exclusively on financial performance as an indication of overall corporate wealth. It was, they said, too short term, remote from most workers' area of direct influence and too easy to fudge. They suggested a fully cascaded approach with three other dimensions added to reports on performance: customer satisfaction, business processes and innovation. They proposed corporate IT systems with these data routinely collected and accessible to all, such that a manager arriving at work could log on, be presented with a single summary display screen showing which of the four areas were underperforming, and then 'drill down' to establish exactly what was going wrong and where. In all probability, the few companies that have applied this approach comprehensively have yielded the kind of returns that Motorola reports for six sigma. In practice, some claim to have adopted the balanced scorecard but simply use it to organize the paragraphs in their annual statement to stakeholders. Others have structured their monthly management meetings into the four areas. A few display charts on the shopfloor in the four areas. However far you take this approach, it provides a focus for the goal-setting process of the organization as a whole.

If nothing else, the balanced scorecard literature reinforces the importance of comprehensive commitment from the top to the process of change and the recognition that money is not the only factor of importance.

Initially, four areas tend to be the focus of attention:

- the customer: reflecting how your customers see you
- financial performance: how you look to your shareholders
- internal business perspective: highlighting what you must excel at
- extent of improvement: or how do you continue to get better?

There is no point in introducing a new culture into the business if it does not produce tangible, measurable benefits. Too many organizations end up three years down the road doing all the right things but achieving little. This is generally because appropriate measures of critical success and short-term goals were not agreed at the outset.

The importance of this step needs stressing. One benefit of doing it thoroughly is that it gives a further opportunity to discuss and clarify the overall vision. Another is that everyone understands and has ownership for the measurements that are eventually agreed. This activity is a natural continuation of the initial clarification of goals.

Self-assessment

With the critical success factors as a framework, most organizations that successfully adapt begin to look carefully at the corporate structure and systems. This should identify any major changes that are necessary to allow the new culture to flourish.

The most common themes on which to focus are:

- customer orientation
- organization structure
- people systems (reward, recognition, appraisal, etc.)
- internal communication.

Most organizations are arranged by function, whereas most business processes, such as invoicing the customer or developing a customer relationship, require input from several functions across the business. For example, are your cost centres set up to reflect a customer need or a financial one?

You will have identified some organizational barriers that exist. Now you need to decide how to address these issues. The changes that many businesses need to make, even the best ones, are often substantial.

If external consultants make the recommendations, then there is very little chance of their being carried out. Instead, a team of people from inside the organization, possibly with an external facilitator, should make the assessments and present recommendations to the senior management team. This means that:

- the activities are owned internally
- the new culture of greater involvement can be practised (and any problems ironed out)
- the level of understanding can be raised around the company
- the senior managers can gain some experience of practising the new culture.

The executive team has to agree its preferences for immediate action, to understand the support needed and its own role in providing this. Working with two or three teams, skilled facilitators will help them to prepare sound recommendations for actions on these priority areas. These teams can be expected to meet for a few hours each week, and for up to three months.

A few organizations on the six sigma trail have bought in to the idea of coloured 'belts' for their facilitators. Based on the *Kyu* grades of eastern martial arts, a facilitator takes a series of courses and practical assignments and progresses up the ladder of respectability. Aside from being an insult to the many years of study and training put into martial arts and its spiritual core, most of the belt training focuses on mechanistic problem solving, with very little emphasis on individual development, interpersonal transactions and group dynamics, which should be the real focus of the facilitator.

Many of these training sessions are offered by 'experienced practitioners' rather than experts in their field. As mentioned elsewhere, several excellent training programmes based on sound principles and depth of understanding of process dynamics, are slowly emerging, such as the MSc programme at the University of Surrey on Change Management, the Diploma programme in OD and change run by the UKCP accredited, psychotherapy training centre, Re.Vision in north-west London, and other programmes at the Roehampton Institute, the Gestalt Centre and the Tavistock Clinic.

Critical success factors and self-assessment at Motorola

So far, the experiences at Motorola had confirmed to them the importance of quality as a competitive asset. While some organizations dictate the critical success factors downwards, at Motorola the first wave of improvement had forced people to re-examine their own work, reassess their customers' needs and put their own measures of performance into place.

The self-assessment step thorough which most companies go provides an opportunity for people, often the most senior managers, to accept for themselves that there really is a need for change. There are many examples of this happening in Motorola during its first improvement phase.

By 1979, Galvin had recognized the importance of constantly upgrading the skills of all of Motorola's employees. He supported a five-year training plan to introduce new technology and reinforce teamwork. This commitment was made despite a less successful initiative in the early 1970s when he established the Motorola Executive Institute to provide up to 400 managers with what he envisaged as an MBA equivalent in only four weeks. (It is surprising how many companies arrogantly assume that they can do this. They rarely ask why the likes of Harvard, Henley and INSEAD insist on year-long full-time programmes. It should perhaps be a salutary warning to prospective MBA students as to how their qualification may be viewed when they re-enter the job market.)

By 1984 executives at Motorola, including Galvin, were questioning whether the pace of change was quick enough. A presentation made by Bill Smith in 1985 confirmed these fears, and sowed the seeds of the six sigma process. An examination of the early-life field reliability of products showed very clearly that if problems and defects needed fixing in manufacture, there was a good chance that the product would go wrong during its early period of use. If defects were designed out before manufacturing, then there was a good likelihood that the product would not break down when used by the customer. The emphasis had to be shifted to better design, not only for manufacturing, but in everything that was done. It was not sufficient to focus on simple improvements.

While, with hindsight, this may seem blindingly obvious, it contained two vital messages: first, that the focus should be primarily on design and only then on implementation, and secondly that it was not merely in manufacturing that problems occurred, but in most business processes.

In 1986, Galvin visited customers and grew impatient with the common concerns that he was told about. Order completion, transaction accuracy and delivery were all areas to which customers alerted him as being inadequate.

When a company launches a quality improvement process, the most senior person with responsibility for quality often becomes a focus for gathering performance statistics, and Motorola was no exception. Richard Buetow, their Senior Vice President with responsibility for quality had gone one step further. Through a sophisticated approach to competitive benchmarking, he had carefully monitored defect levels among Motorola's Far East competitors. While Motorola was one of the best American companies, with defect rates of about 6000 parts per million (ppm), he found that Japanese competitors with the same equipment were achieving less than 4 ppm.

Motorola's constant re-examination, and its refreshing lack of complacency, make it all the more remarkable that it achieved the US Malcolm Baldrige National Quality Award in 1988. The issue was not with the award or Motorola's suitability for it. The problem was that by 1987, Galvin and his team had already realized that what they had been doing was merely scratching the surface. It is simply that what to Motorola was 'scratching', represented major earthworks for other companies, not because they were any worse, but because Motorola's goals were much more ambitious.

Ironically, the final trigger to change came in the year that they were awarded the Baldrige Award. Their RISC chip was months late to market because of design problems, a quality problem that represented a major market cost to Motorola.

Activity

Have a word with someone in the purchasing department of your organization and draw up a list of the different firms of consultants that have been engaged by your company in the past ten years. See whether you can obtain copies of the final reports that these firms produced (most consultancies do). What was the range of topics studied?

There may be an opportunity for a quick win. Are any of these worth revisiting?

Perform a simple critical success factor exercise for yourself. Speak to one or two more senior staff, and try to ascertain what you would need to do to be sure of promotion at the next round. In other words, what are *your* critical success factors?

Cascade the aims

An organization may be thought of as consisting of executives, middle or supervisory management and non-managerial employees. Often the executives will have spent much time, quite rightly, getting to grips with the fundamentals of the new culture and deciding how to fire up the organization. The mistake is then made of cascading the message to the middle management, and then on to the non-managerial staff without sufficient thought for the consequences.

Non-managerial employees are usually quickly excited by the vision, because they are going to be listened to more, trusted more and given more ownership – all important to culture change. They look to their manager for help and support, in the way that they have been led to expect, and nothing happens. End of excitement. Beginning of disillusionment. This is usually because the manager, who has perhaps the greatest behavioural change to make, has not had sufficient input into the process to have ownership of it. Neither has the manager been given the technical or management skills to support the change. Following on, therefore, and developing from the senior management induction, middle management must be brought into the picture. Ideally, they need as much investment in training as, if not more than, the executives.

An important balance needs to be struck in large organizations between internally delivered development (for economy and to demonstrate commitment) and externally delivered development (for credibility and to demonstrate commitment).

The result must be that the managers are as enthusiastic about the change as are the executives, which will in turn ensure that their behaviour does change. As people are much more aware of what their managers do than what they say, this will have an impact on other employees. The cascading process can then, in a properly supported way involving the managers, be carried on to the front-line employees.

Develop skills

Some priorities need to be established for developing people's skills. They will focus on:

- process skills, including understanding individual behaviour, values and attitudes; interpersonal communication skills (understanding the dynamics of transactions between individuals and using these in coaching and counselling); group dynamics (revealed in the handling of teams and leading them to more productive performance)
- technical skills, such as problem solving, systems analysis and quality function deployment, experimental design (e.g. Taguchi's techniques) and statistical process control
- interpersonal skills for employees, including teamworking and customer care.

As mentioned earlier, the initial emphasis should be on the process skills of the manager, and the first managers to experience this should be the members of the executive team.

Organizations around the world report that the culture change only began when they prepared their management teams. This preparation usually involved enhancing their interpersonal skills. By allowing them to delegate virtually all of their day-to-day activities, they can devote their time to developing their employees.

This development makes the executives much more aware of the real mechanism by which a new culture is put into place.

Cascading the aims and developing skills at Motorola

Throughout the late 1980s and early 1990s, the senior team at Motorola reinforced its vision by refining the goals rather than changing the overall strategy. There are few management teams elsewhere who are insightful enough to do this; they usually prefer to rubbish the previous leader's vision and impose their own. This can almost become cultural itself: in one organization in which I spent some time working with the senior management team, it had become a cultural expectation that people were appointed into a new role to 'turn it around', to 'restore confidence in our offer', and so on. The individuals never stopped to question why their predecessors were still employed by the company and had been given larger and more influential roles if they were that incompetent.

By the mid-1980s, Motorola's vision and goals were encapsulated, literally, in a plastic card carried by every employee. It spelt out the fundamental objective of total customer satisfaction (TCS), while on the reverse were the key beliefs, goals and the current set of key initiatives. Armed with this information, no Motorolan had any difficulty weighing up a decision.

Then in 1987, the goal posts were brought closer together. The target remained the same but the objective now was to achieve a further ten-fold improvement in quality in two years, and 100-fold in four years (1992). This last goal was one of six sigma.

For the record, George Fisher, then Chairman and CEO, reported that overall Motorola had achieved 5.3 sigma by the third quarter of 1991. One good example of a non-production improvement was the month-end financial procedure. In a typical month they would handle over two million transactions in their general ledger. Each of these was a chance to make a mistake. By February 1991, they had reduced their error rate to 0.08%: a sixteen-fold improvement in two and a half years. In terms of time saved, the company was able to close its books in four days rather than eight. This represented 576 000 person-hours, or $20 million, saved each year.

Developing skills

Despite the false start in the 1970s, Galvin and his team were now more convinced than ever of the importance of a highly skilled workforce. While formerly training had been provided but was effectively voluntary, they now demanded that every employee should undergo at least five days of training each year. Much of this was provided in-house by the Motorola Training Centre, founded in 1981.

In 1989, the purpose of the centre was reframed and Motorola University was founded. As Galvin pointed out, 'training centre' sounds a lot less impressive than 'university'. Motorola were pioneers of the concept of the corporate university. Since then many companies have followed suit and today some even award degrees through affiliation with established awarding bodies. In the USA their training facilities are superb, and courses are provided by individuals who are carefully screened, selected and coached. The company provides its own syllabus, materials and even terminology to ensure that the message is totally consistent, not only for a given course, but also with all others in their programme.

In Illinois they use staff from the Northwestern University's Kellogg School of Management, in France they work with the professors of the Université de Technologie de Compiegne, in Macao with Asia Pacific International University, and in the UK with the University of Edinburgh.

The relationship with these universities is a fascinating one, and one that other universities would do well to heed. Motorola sees itself very clearly as a customer. It deals with the universities just as it would with any supplier, precisely spelling out its requirements, in terms not only of the quality of the product, but also of the process that leads to its provision.

By the early 1990s, Motorola had recognized the value of rigorous quality control of higher education. For too long, universities had been their own arbiters of quality and, as there was generally no link between the institution and the future employers of its products (the graduates), clearly there was a strong probability that graduates would not meet Motorola's expectations. So, by 1992, Motorola had invited other institutions to attend the Motorola University, the stipulation being that they had to adopt the total quality process for their own administrative systems. Reported in *Fortune* magazine, Michael Cummins of the University of Miami observed that Motorola was saying that if universities did not use these techniques to improve, then their graduates will become too expensive to employ.

By 1992, Motorola's education programme was costing some $100 million, but independent auditors had shown that the return was in the order of 33 to 1.

One of the concerns that Galvin's team had before launching the six sigma initiative was that many of the line workers lacked even the basic skills of the 3 Rs: Reading, wRiting and aRithmetic. Their realization was a decade ahead of that of the US government. A study by the National Center on Education and the Economy warned of the dilemma facing the US administration: many American workers continue to be in need of these basic skills; however, the skills are not called upon in their day-to-day lives, so there is no incentive to learn. So long as American companies continue to use traditional management control to organize their work, this will continue to be the case. But this approach only works while employment is falling and technology is standing still. At the time of writing, employment rates are at their highest levels in twenty years and technology continues to escalate: basic education has never been so crucial.

The prospectus from the Motorola University reads like a menu of the top engineering subjects from the best colleges. There are even courses that many universities have not yet been able to develop and run. There is also a comprehensive range of management skills courses as well as training in simple problem-solving tools and techniques. Essentially the programme comprises relationship skills, technical skills and business skills. Again, a decade or more ahead of its fellow countrymen, in 1992 Motorola committed $5 million to providing remedial education for its line staff.

Supporting teams

A cornerstone of successful contemporary business cultures is that people naturally work in teams. Probably one of the least recognized, yet most significant, changes in the organization of work over the past decade has been the way in which team-based working has become endemic. Even in the 1990s, we struggled to introduce the ideas of job sharing, multitasking and other collaborative approaches. I cringe to remember the days of quality circles and self-managed or self-directed teams. Today only a few bastions exist where an individual works in isolation. That said, this is true for corporates but we also have to recognize that today we have the largest proportion of self-employed workers than at any time since the First World War. What has happened is a polarization. Perhaps it is time that the education system acknowledged this and prepared individuals accordingly.

Organizations pursuing six sigma today need to examine this current state of development of teams and how much further they can be implemented. The essential message is that the six sigma climate depends on groups of people working together at the point of delivery. Effectively, they are making all the decisions that they can that affect their work and their motivation to deliver it. There have been countless experiments in the past decade ranging from co-operatives, collaboratives, common studios and entirely self-employed, to self-directed. Whereas ten years ago we might have prescribed the best approach, today it is more a question of experimentation. This is another reason why following 'experienced practitioners', whose methodology is probably based on the early work at Motorola and General Motors, and not in today's working climate may be inappropriate.

If you run a manufacturing business, for example, do not assume that everyone who works in your factory has to be employed by you. Whether a person works as a cleaner, a machinist, quality control technician or supervisor, each has a choice as to whether they work for you or are a contractor to you. The key is that you have to enable them as a team rather than managing them as a bunch of employees.

In your pursuit of six sigma, the executive team has some serious decisions to make about the organization of the workforce.

Even mainstream employers such as the NHS have to embrace this dramatic shift in the definition of employment. The NHS in the UK is the third largest employer in the world. (Number one is the Chinese red army and number two is the Indian railways.) Yet the team that delivers care to a patient in a hospital may include 50 per cent who are self-employed and the rest may all work for different employers. The picture is complex.

Supporting teams at Motorola

The power of teams has been a part of the Motorola culture for many years. Whether you look to the East, at the power of the consensus style of Japanese businesses working within a well-defined hierarchy, or at the self-managing work teams in some Western businesses, most notably Johnsonville Sausage, by the 1990s Motorola had not embraced the concept as fully as these companies. At Johnsonville, for example, employees permanently worked in teams, differences in status were virtually absent and individuals were so multiskilled that specialization no longer remains (Stayer, 1990). No doubt, if it is worth trying Motorola will progressively experiment with this shift in the next few years.

Self-managing teams are seen by most organizational development specialists as a natural consequence of empowerment cultures. When establishing them, the majority of companies began with teams made up of individuals from different areas working on topics identified by the management group: so-called 'task forces'. They were then institutionalized as self-managing teams. Two initiatives that nonetheless moved Motorola in this direction were their participative management programme and TCS teams.

Motorola introduced an international competition for its team-based employees. Often as a part of their day-to-day work, the teams choose a problem and then set about using a simple six-step problem-solving process to resolve it.

This problem-solving process if applied by individuals would, of course, lack the power of group process to improve its effectiveness. It was a normal divergent–convergent thinking model of the type that has been widely described since the late 1950s. Variants of these models are taught on management and problem-solving courses around the world. Nevertheless, the dramatic results that are achieved by teams that use them are a good indication of how poorly they have actually been applied.

If they wished to, the Motorola teams, known internally as TCS teams, could present their findings to panels of managers and, since 1993, customers. They were judged on criteria that included project selection, teamworking, the analytical tools they used, the evaluation of alternative solutions, quantifiable results, the extent to which these have been communicated and adopted elsewhere, and the team's presentation itself. Winners at various levels progressed to national, regional and international competitions. The results were finally included in the company's annual report. By the early 1990s, there were about 4000 TCS teams representing nearly 24 000 employees, or 25 per cent of the workforce of 102 000 people. About 50 per cent of the teams joined this competition.

Most authorities would suggest that creating internal competition is likely to undermine a culture change process in the long term.

Perhaps understandably for an organization speaking engineering, Motorola's approach put little emphasis on the group dynamic facilitator, and focused heavily on the tools and techniques of problem solving. Even then, their approach concentrated on analytical tools and techniques rather than intuitive ones.

Counselling members

The key to the transition to greater involvement, which is what the Motorola programme really sought, to taking full responsibility and authority, lies in the attitude and behaviour of the managers. Anyone asked to change from being responsible for something to handing over that responsibility to someone else finds it difficult to accept. It is particularly hard to swallow when that person was previously regarded as several steps lower down in the hierarchy.

The managers need to adjust, not only to cope with this loss of personal power (a palpable grieving process), but also to change their management style from that of 'controller-cop' to 'developer of people'. To help them to do so, most successful change processes nowadays include off-line coaching for the management team.

Similar provisions are made for staff. Some companies, for example, will be offered facilitator-based coaching, others have individual corporate coaches who move around among the different locations, some train up experienced staff to take on a development and support role. Many will continue to ignore this. A surprising number of individuals (especially from non-managerial levels) make use of the counselling services provided through the employee 'assistance' schemes to obtain the kind of help that their company could or should be offering.

Targeted marketing

Throughout the transition process organizations need to develop plans for a targeted approach to marketing the new culture. This begins internally, but eventually includes external marketing. Initially, it is intended to create a demand for the new culture, and later to explain the detail of the process. The external phase involves suppliers and then customers.

Within Motorola, at the management level, articles, videos, booklets, books and courses all reinforced the message of six sigma. While some companies expend a great deal of energy on this process, at Motorola it was part of the culture and had probably been so for a long time. It was this openness to discuss new ideas that had stimulated the change in the first place.

We have already described the card in the pocket reinforcing the goals of the business. The TCS teams were exposed to corporate messages and a variety of materials was provided to plant managers for general distribution, including the library of corporate philosophy already described.

For many office-based employees, the existing corporate culture is again the key to the internal marketing process. Informal and formal networks exist throughout. While some companies are afraid to use the grapevine, Motorolans and other savvy companies use it to advantage. While a formal organizational structure may exist, it is widely superimposed on a matrix of responsibilities.

Ongoing review

Finally, one step that it is easy to forget, and yet one at which Motorola excelled. It is very clear to anyone reading the history of Motorola's six sigma drive that this was a topic that was constantly on the agenda of every management meeting. As soon as it drops off, you see the beginning of the end: the staff soon realize that this is something that is optional, that managers are paying lip-service to.

By constantly reviewing the successes (and promoting them) and addressing any shortcomings the executive team sustains the initiative.

References

Kaplan, R. S. and Norton, D. P. (1992) The balanced scorecard – measures that drive performance. *Harvard Business Review* 70(1): 71–79.

Petrakis, H. M. (1991) *The Founder's Touch: The Life of Paul Galvin of Motorola*. Chicago, IL: Motorola University Press.

Stayer, R. (1990) How I learned to let my workers lead. *Harvard Business Review* 68(6): 66–83.

Wilson, G. B. (1993) *Making Change Happen*. London: FT Pitman.

4

The product development cycle

Some people try to distinguish between new product development (NPD) and the evolutionary development of existing products. The problems that are encountered when introducing a new product, and during a major innovation to an existing line, are actually very similar. In many ways the two should be seen as different points (not even as extremes) on a continuum. At one end is the step-like evolution of punctuated equilibria and at the other a slow development of features or phyletic gradualism (Figure 4.1) (Cracraft and Eldredge, 1979).

In the punctuated equilibrium model a product is launched with its particular set of characteristics. It competes in the marketplace for a period of time. Not all of its features may be unique, and some may not be exploited. Then the marketplace changes and the presence of certain features may be more or less important. New competitors appear with these features while previous competitors who lacked them disappear. This leads to a pattern over time of steady products periodically shaken up before settling on a new format which remain consistent for a while. We see this particularly when a major innovation appears in a product group. For example, mobile phones have been through several such phases. At first these were associated with reducing size, then with the introduction of feature sets such as downloaded ringtones, more recently with image capture and now with streaming video. Each time, a whole new generation of phones appears in a short period.

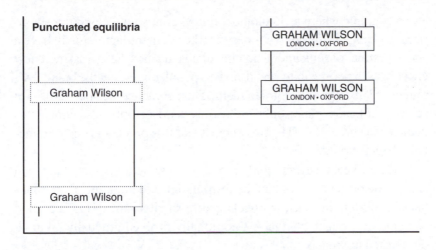

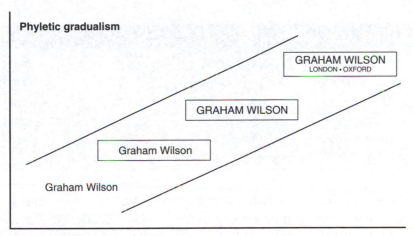

Figure 4.1 Punctuated equilibria versus phyletic gradualism

The phyletic gradualism model says that features are slowly enhanced and dropped, yielding a slow and progressive change to the product. Sony's Walkman range, and the personal stereo market generally, is a good example. The initial products were small cassette players. While manufacturing quality has slowly improved, it is still possible to buy basic machines that are essentially identical to those of twenty years ago.

These two models of product development are not mutually exclusive, however. While basic portable cassette players can still be bought, the next generation emerged as CD players, the next as MP3 players, then DVD players, and so on. Each new generation appears, but the previous one continues to evolve, at least for quite some time afterwards.

An example where technological innovation is not behind the evolutionary change is the confectionery market. A manufacturer may have a stable product of milk and plain chocolates in a box. They may add new chocolates experimentally and drop less popular ones, but the brand, 'X', remains. This is phyletic gradualism. Then someone decides to offer a selection of only plain, only milk, only hard or only soft chocolates launched as brand 'Y'. The two coexist and this process is one of punctuated equilibrium.

In recent years, several confectionery manufacturers have launched single-bite products, based on the formulation and branding of an established multibite product, but packaged in smaller units. This is a good example of products that have traditionally evolved gradually taking a punctuated approach.

Activity

Spare a few minutes to consider your own products or services. Try to find examples of the two models happening.

One of the tricks of strategists, when looking at product development, is to see whether they can spot ways in which a product stream that has evolved mainly through phyletic gradualism can be boosted forwards by means of punctuated equilibria.

One firm making steel structures conducted this historic review and saw that their products had only really ever evolved gradually. They asked themselves what it would be like to evolve in 'fits and starts'. The conclusion was to look for other people's products that they could acquire that achieved the same end result as their own but did so in innovative ways. Within twelve months they were selling none of their originals, but exclusively sold the new product. There was a lot of change associated with this shift, much of it painful, but it has ensured the survival of a business that was otherwise doomed.

Most 'new' products occur by developing an existing one. Even quite extraordinary innovations grow in this way. For example, Edison's original electric light bulbs, which replaced gaslights, were developed for safer illumination aboard ships. They were not dramatically different in structure, and used the same fittings. Nor was the distribution of the power very different: the wires were often even fed through the gas pipes. The quantum leap to the next application of electric light is equally typical, however. A fanatic supported the development of floodlighting for night baseball: only then did the concept of electrical street and domestic lighting take off.

One useful way of looking at the design activity is to distinguish between 'product design' and 'process design'. Product design starts with a basic idea of a product and then evolves that idea. Process design tends to start one step earlier by asking what it is we are trying to do and thence what is the most effective way of doing it. Product design is much more common, but process design is what sets designers like Philippe Starck apart from others. He begins with a premise: 'I want to extract the juice from an orange'. Juice extraction is the process and then he finds the most effective way of achieving this. More mundane designers will say, 'We have this juicer, now what can we do to improve it?'

Product design

When we develop something from scratch we often begin by defining what the product is. This product design step may consist of only a thumbnail sketch or it may be much more sophisticated, even at this stage. However, when we are enhancing an existing product, this step is often neglected. When it is included we can get quite surprising new ideas.

Product design is the stuff of history books. Take the 'simple' toaster. There are two problems associated with toasting bread. First, we want to be able to toast both sides, and secondly, once the toast is done we want to be able to butter it before it becomes too cold. For many years, toasters consisted of a pyramid-like structure that sat above a gas flame. They toasted only one side at a time and the toast and the toaster became very hot, making it hard to turn the toast and do the other side without burning the user. Toasters of this style are still sold through high-street stores.

The next step in the evolution was to have side panels that could drop down and as they did so flip the toast over, allowing us to brown both sides without having to touch the toast. Toasters of this style are also still sold through high-street stores. Eventually, the source of heat shifted from gas to electricity, although the toasters often looked very similar.

Inevitably, though, the toast eventually still had to be removed from the area of the heating elements, without burning the user.

Product design redefined the device. Inverting the toaster, by doubling up the elements and moving them to the outside of the toast, meant that both sides were done at once. This brought the added advantage of halving the toasting time. But it also brought disadvantages: the fingers, when retrieving the toast, were dangerously close to the red-hot electrical elements.

The next step was to provide a simple lever that could be used to eject the toast without burning the user. Again, devices of this design are still successfully sold in high-street stores. Often such toasters had a totally separate clockwork timer added. It was simple evolution to link this to the lever, and the 'pop-up' toaster was born.

Process design

Over the years there have been a few forays into the world of process design for toasting bread. In the 1970s, strange tong-like devices were made. The bread was buttered and placed into the tongs, butter side out. The filling was placed on one dry side of the bread, the tongs were closed and clamped together and the device was put over a gas flame or an electric ring. You can still buy devices like this.

The (toasted) 'Sandwich Maker' came on the scene in the 1980s: the same process but this time with built-in electrical elements and 'shell-like' clamps giving pretty sandwiches. Eventually, in the 1990s, Kelloggs used process design to innovate. Their 'Pop-Ups' contained a filling in the middle of a sealed envelope of pastry, hence there was no need to butter them at all.

Looking at the microwavable foods available today, we can see many more examples of process design radically changing the way in which foods are prepared and served.

Process design is not restricted to the manufacturing sector. For instance, the concept of providing a banking service is common to most countries. Anthropologists can identify its origins in even some of the most primitive societies. And yet, despite this, or perhaps because of it, the service differs tremendously from country to country.

Sometimes the system is the same, but even then there are significant differences. The British were largely responsible for introducing many of their former colonies to the concept and practices of modern banking, but go into a bank in New Delhi and it will bear little resemblance to its

British counterpart. Banks in France are radically different from those in Britain. The product is the same, but the process is different.

Even the use of modern technology changes across frontiers. Banks in some Gulf Region states, for instance, have traditionally provided an air-conditioned meeting space for business people. Until quite recently they operated dual queuing: one line for withdrawals and the other for payments. Even their cash machines had to be installed differently. Because of the importance of providing space for meetings with air-conditioning, most have been installed within the buildings, but to do so means that the process by which they are used is different from that of their European counterparts.

By distinguishing between the product design (what it is) and the process design (how it is used), we can begin to unlock enormous possibilities for NPD.

It is often difficult to see why there needs to be a distinction between product design and process design, especially when we are looking at a product or service that we know intimately. The following two exercises, which come from workshops on creativity for businesses, may help in clarifying this distinction.

Activity: product design

From studies of humming birds and other animals, it has recently been shown that people can assess the urgency of danger from the dilation speed of their pupils. This can be simulated by a bright object growing in size.

On most pedestrian crossings there is a light-based sign to encourage pedestrians to cross, to finish crossing quickly, or not to cross at all. The sign itself varies from country to country. In the UK it is a symbol of a man in green with his legs apart, or in red with the legs together. In France there are lower level lights with just a red or green colour. In the USA the signs are in white saying 'Walk' or orange saying 'Don't Walk'.

Try redesigning the sign to make use of the new knowledge on pupil dilation.

In the 1980s and 1990s there was a boom in the 'personal organizer' market. At first, these organizers were simple filing systems, some quite well established, such as the Filofax. Newer entrants, such as the Time Manager, offered not only the diary but also an extensive range of training options in how to use it.

Then electronic organizers such as the Psion took off. Subsequently we have experienced a massive growth in personal digital assistants (PDAs). So far, I am not aware of anyone offering dedicated training for a particular PDA. This is probably because of the consistency of the software with other Microsoft products, but the potential exists.

Essentially, the sort of product that you use determines, to a greater or lesser extent, how you use it.

Activity: process design

Why not re-examine your existing product range? Look beyond what the product or service is to how it is used. How does this way of using it differ from the way in which competitors' products are used?

Could anything be done to your products or services to enhance, or even create, this difference?

An excellent example of this happening was the emergence of Daewoo's car showrooms in the early 1990s. They chose to differentiate themselves not on the basis of their cars, but on that of the sales process.

What happens when things go wrong?

There is a well-known car rental company that undoubtedly outperforms its competitors in most of its operations. However, if you are a disgruntled customer (one of the few who bother to complain) you will be shunted from pillar to post, and when you eventually fall into the hands of the customer service department you will be treated like a criminal until they have verified the facts of your case. They are, I am sure, very fair. They assess the value of your complaint and credit your charge card accordingly.

If you are exceptionally lucky they will also write to you to apologize, sending a standard letter, drawn, you feel, from a repertoire of stock paragraphs stored as a Microsoft Word template on the company network. If it was dictated, then the script would go something like: 'Send Mr ... of ... paragraphs 1, 33, 12 and 7. The amount in paragraph 7 is £33.24'. For reasons that are far too unclear, sometimes the letter conveniently falls off the pile and you get nothing, except for the charge card credit.

Alternatively, take the equally well-known manufacturer of industrial radio-paging systems. Most of its products are designed for the individual customer. A salesperson, based at the head office, takes the order. The details, on a complex ordering form, are sent to the design department. The design engineers are housed in a pleasant out-of-town location in a 'silicon valley' at least 100 miles away. They create work instructions from the designs and pass them to the production department. For reasons of costs, the production people are based in a cheaper part of the country. This has helped to keep overheads low and so improved the company's competitive edge.

Once the system has been installed and you, the customer, have signed for it you pass into the hands of the after-sales service department. When problems arise, you telephone them. You enter a strictly administered holding system so that when your 'ticket' comes up an engineer will be despatched to you. The engineer arrives, fixes the problem, and completes a call-out sheet describing the nature of the problem, how it was fixed and how long it took to fix. You are asked to sign the form. There is no reason why you should be asked to, given that the service is covered under the maintenance contract that you automatically have. There is also no reason to sign because once the forms have been returned to the after-sales service department they are simply stored in a large boxfile system under your name.

This department is based back at head office. This is not very practical because most customers are outside the city, and the engineers have to struggle with the traffic at least once a day. This also, very effectively, prevents the designers learning from previous models, and has led to a downward spiral of non-communication as each department blames the other for the poor reliability of the products.

It is very easy to be critical about such organizations. Unfortunately, they are typical. There are countless very good 'excuses' why they have developed that way. Often, whole industries are like this.

Compare these two examples with the case of the Joban Hawaiian Centre in Iwaki, Japan, about 200 km north-east of Tokyo (Yabe, 1990). The Centre is a theme park, built on the site of an old coalmine, which received the coveted Deming Prize for Quality in 1988. Its trigger to change came from the falling attendance figures in what had once been a highly successful operation. In true total quality fashion, employees were coached by their managers as they identified sources of poor quality, as perceived by customers, and then implemented improvements to redress the balance.

For example, plates of food cleared by waitresses were taken into the kitchen and, before they were washed up, the remains were scored on wall-mounted tally sheets. Among the results of their analyses they discovered that children preferred consistent quantities of spicy sauce with their rice, and smaller portions of meat dishes.

Another attraction at the resort is hula-hula dancing. Here a quality circle was formed. Unlike its British counterparts, this was a genuine quality circle. The employees chose to come together themselves, rather than being directed to or selected. They chose their own topics for improvement, too. The hula-hula dancers noted variations in each other's performances. By counselling one another they were able to resolve many problems. Other issues involved hard data. For instance, some women were spending as much as £75 per month on make-up, while others spent less than £20. This inconsistency seemed illogical to the dancers, and they set about tackling it. Among the improvements that they introduced was a standard make-up procedure, and each woman now receives grooming tuition as part of her induction training. Since then, the dancers have all been involved in peer assessment of each other's performance and appearance.

Throughout the process, customers' views are sought, carefully quantified and analysed. Then customers and staff work together to devise improvements not only in the product, but also in the systems that led to it.

Hearing the voice of the customer

For years the gurus have been telling us that the only way to improve quality is to listen to our customers, and yet time and time again, as our car rental company and pager manufacturer show, we do no such thing. It seems remarkable, but very few quality improvement or customer service initiatives actually consult the external customer within their first two years. This aspect of the pursuit of six sigma is so important that another activity from our creativity workshops is included here.

Activity

Conduct a simple audit of your own organization's efforts at improving quality or customer service. What process was followed? To what extent were customers really consulted? How were solutions developed? How effective have they been? We would be very surprised if many could show a direct connection between customer input, change in the organization and a subsequent improvement in customer satisfaction.

What opportunities do you provide for your customers to be heard in the normal course of business?

Have you, or a member of your family, tried the product or service for yourself?

We have all heard examples of complaint boxes with sealed lids or no pen. The scourge of the internet seems to be standard replies to common complaints and a very complex route for anyone to make contact with the company. (Try sending an e-mail to a real person at the manufacturer of your computer or to the maker of your software.) Are any of these situations true of your own organization?

Look at your competitors. Are there any differences? For instance, for a long time, Texaco was one of the few petrol companies in the UK to advertise an 0800 freephone customer service number. Why do you think they were doing so? Most of the calls were about their loyalty programme.

If, when you've finished, you still feel that things are pretty good, then ask a friend, someone who you know will be honest, to do the same exercise. Then listen.

It is common for senior managers to be embarrassed by this exercise. They often find themselves defending their organization. Most organizations do not have any listening devices for customers. It is hardly surprising, then, that the few customers that they do hear from are those who go to extremes, and are therefore easily branded as cranks. This creates a self-reinforcing sense of security for the company.

Do not fool yourself just because your organization has a large number of front-line staff and lots of apparent customer contact. As Jan

Carlzon of SAS is well known for pointing out, even if they are heard, most of these 'moments of truth' are lost forever in the organization's internal maze.

This is because even if they do listen to their customers, many organizations have real difficulty overcoming their own internal prejudices, bureaucracy and functional barriers to allow the problems and ideas to move towards the creators and upstream deliverers of the service.

The concept behind the product development cycle is to recognize that this flow is important, and to provide the tools to allow it to work effectively.

Why don't the problems get resolved?

In just the same way that the domestic light bulb only took off because of a dedicated baseball fan, when problems begin to recur usually someone senior (who is growing fed up of hearing the same message time after time) will eventually take the initiative and act as a kind of internal champion. This is true of new ideas as well as regular complaints. Wherever they have come from, it is not until a senior champion takes them on that anything will happen.

For most companies, the interface with the customer is the salesforce. If enough customers complain and enough salespeople bother to relay these complaints to their manager, then they may be dealt with. This is pretty rare, but it is even rarer for a customer suggestion to be forwarded through this route.

If the sales manager is not too busy, and if he or she has the opportunity, then they may relay this information to the sales director.

The sales director has the same obstacles to overcome, but may get around to telling either the marketing director or the operations director. If the marketing director hears about it, then he or she has a range of ways of tackling the problem. If the operations director hears about it, then he or she will also have a range of solutions, though probably different ones. (In the case of a problem, the easy option is to blame one another, but they seldom do this.) Regardless of who picks up the problem, the route that it takes is up-the-ladder and down-the-ladder, with lots of filters in each direction.

Whether the need to develop the product comes from the outside customer, market trends or an individual entrepreneur, the need and the solution will be subject to the same filters and passing on, just as a regular complaint is. If the people along this chain do not share the same offices, as was the case at the paging company, then serious breakdowns in communication can occur.

The Royal Mail provides a classic example of this. To post a letter people need stamps. Because stamps are prone to running out, the Post Office provides convenient machines selling books of stamps, usually close to main post offices. A small provincial town has three such machines, one by the main post office, one by the sorting office and the third outside a subpost office. Responsibility for the first box belongs to Post Office Counters Ltd. Responsibility for the second lies with Royal Mail Letters Ltd. No one seems to know who is responsible for the third machine.

Normally there would be no problem with this arrangement, but what happens when all three machines run out? As a business they risk losing customers, customers who do not distinguish between the elements of their internal organizational structure, customers who have an important letter and just want a stamp. You could dismiss this example as petty, and in some ways it is. But it is worth putting 'petty' examples into the context of six sigma. In this particular case, if the machines are empty only one night in a year, the Royal Mail has a 99.7 per cent availability rate. This represents 2739 defects in a million, a far cry from the 3.4 defects per million that we are striving to better through six sigma, and that organizations such as Motorola achieve.

It ain't what you do, it's the way that you do it

Whether it is the production or operations director who picks up the problem, the solution is driven by the people who report through him or her. This often means that someone will be given the job of investigating it and recommending changes. Even in very large, highly sophisticated businesses, the tools and techniques that they will use for this analysis will be quite simple.

Unfortunately, I have often seen quite senior engineers and graduate chemists put forward completely unsubstantiated plans. This can happen for a variety of reasons, although the most common is that they just do not understand what they could have done to resolve the problem.

It is a production problem

If the production people do tackle the problem systematically, then in relatively sophisticated companies they are likely to draw on a combination of two techniques to help them. The first is statistical process control (SPC). The second, and much rarer, will be some form of experimental design tool.

It is the designer's problem

If the marketing director is the person who responds, and assuming that they are not going to try to pass the buck, then they will often appoint someone with a design interest to review the design. Until fairly recently there were few systematic tools for them to use. Rarely will they draw on some form of market research. They are much more likely to brainstorm ideas internally, then brief designers and let them get on with it.

People who tend to gravitate towards jobs in different parts of a business often have very different types of interest and personality. Many sales-driven organizations will have a lot of fairly gregarious employees who enjoy the company of others. They gain their energy and inspiration from other people, which is vital for their particular job, but it does not encourage systematic problem solving. They would rather be out partying, or at least involved in high-adrenaline meetings, than collecting and analysing data.

Regardless of the people who try to tackle the problem, there are very few tools available to help translate customers' actual requirements into a designer's language, and then turn the designer's ideas into production requirements.

One technique that emerged in Japan in the 1980s helps us to use the information gleaned from our customers, and so define the qualities of our products or services. The technique is called quality function deployment (QFD).

The product development cycle

The product development cycle is a simple idea. It recognizes that most new ideas are simply reiterations of old ones. It also recognizes that there are three discrete steps in designing a new product or enhancing an old one. The cycle links three of the major techniques used in quality improvement into one formidable weapon. It does not claim to be the only use of these tools, but it is certainly a highly effective one.

In the past two decades, as people have become much more aware of quality, they have tended to latch on to particular elements that interest them. Part of the Lucas group, for example, became the UK's foremost practitioners of Taguchi's experimental design methods. This was mainly because the chief executive was so fascinated by them that he heavily endorsed their use. Such was his enthusiasm that almost everyone in the organization learnt to use the techniques. Most businesses would not dream of teaching their general labourers or cleaners how to devise experiments, yet they did so at Lucas, and with tremendous results.

In some organizations the tool of choice has been SPC, in others benchmarking, cost of quality assessment, and so on. For every tool or technique there will be an organization somewhere that adopts it almost fanatically, to the detriment of other approaches. Our simple product life-cycle model assumes that most tools have a place somewhere in the improvement process, but SPC, QFD and Taguchi's techniques, in particular, will be used in the pursuit of six sigma. The approach can be applied whether you are concerned with product design or process design, with a number of common steps that you will go through. The cycle is illustrated in Figure 4.2.

System design

The basic image of the product emerges in the system design stage. This is usually where its new selling features are identified. 'Wouldn't it be good if it could do this or that?' is probably the most common question at this stage.

This is the area that most people associate with NPD. Certainly it is the easiest for outsiders to contribute to. It is also usually the most fun, and the best time to involve your customers and suppliers, too.

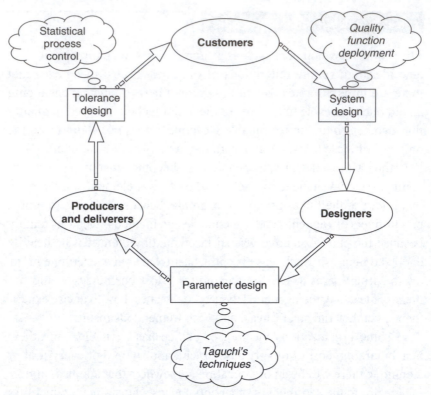

Figure 4.2 The product development cycle

The British Broadcasting Corporation (BBC) reports that people often write to programme producers with an idea for a new programme, usually in the form of a very short description. 'Why don't you … do a programme about badgers in Bristol where you track the occupants of a sett for a couple of weeks and just show the programme live at 11 p.m. each evening?' Then, when the programme is shown a couple of months later, they write demanding their royalties. They do not realize the enormous lead times to which the BBC operates. The system design stage probably happened at least a year, if not two or three years, beforehand.

The system design stage also provides the opportunity for widespread employee involvement. Unfortunately, in many traditional companies this stage is sorely neglected. Think of the whole design process as a progressive filtering and refining of ideas. It should be clear very soon that the more ideas put in at this stage the more chances you have of succeeding.

The success of the system design stage depends on the culture within the business. If it is acceptable to ask questions, to show an interest, to innovate, then more ideas will flow. If people work in teams and are genuinely empowered, then they will be constantly building on simple ideas and turning them into new winning formulae. With the right style of leadership – one that nurtures the employees, developing their problem-solving skills, helping them to get more from one another and encouraging them to take bolder steps – the system design process will thrive. Chapter 5 describes this culture, and why it is so important to the whole six sigma process. Without it you cannot expect to achieve even basic quality, let alone six sigma.

Often the system design stage can be the first highlight of success for organizations. As David Fanthorpe of Black & Decker commented early in his organization's improvement drive, 'We have set free prodigious amounts of previously repressed energy'.

The success of many Japanese-style suggestion schemes is due to their openness to suggestions at this stage. In Western schemes, ideas are often valued according to their level of complexity. One large catering organ-ization recently reviewed the results of a major employee suggestion process. Even though it was intended as a significant motivator to staff and a serious initiative to improve corporate performance, of over 1000 employee suggestions only a handful were given more than a cursory 'thank you', and many were not even given that.

Workers who feel that their ideas are only 'simple' ones are inhibited from contributing.

All the simpler problem-solving tools can come into the fore in the system design stage. Brainstorming, cause-and-effect charting and Metaplan all have their place.

Although the days of using the back of a cigarette packet as a jotter are long gone, the system design stage is very much concerned with this type of innovation. Remember, however, that it need not be restricted to 'why not make this?' Instead, look out for 'why not make it this *way*?' or 'why not *use* it like this?'

Estimates vary considerably, but some authorities have suggested that Western engineers spend about 70 per cent of their time involved in system design, whereas their Japanese counterparts spend only 40 per cent (Ealey, 1988). This is not a contradiction of what we have already said about the Japanese suggestion schemes. The effort is probably much

Table 4.1   Getting the most from suggestions

Demystify the suggestion scheme. Make it incredibly easy for people to contribute

If you have made the mistake of linking the suggestion scheme to some kind of reward, then err on the side of generosity

Obtain the maximum mileage out of every idea. Even if you do not use it, make sure that everyone knows who contributed it, and how much you appreciate their help. Do not just broadcast this once, recycle it at least later that quarter and at the end of the year. Keep it hot! It will stimulate more (maybe better) ideas

Consider an 'ideas' festival. Several companies have developed very effective month-long, short, sharp and fun programmes to start the ball rolling. These invariably pay for themselves

Be innovative about how you recognize people's achievements. Is there someone in-house who has a creative talent: art, photography, pottery, sculpting or whatever? Ask them to design something. Sponsor a theatre trip or a sporting event and then give your people tickets. The possibilities are endless

more effective in Japan, where the engineers are less involved in system design than they are in the subsequent stages of the design process. System design is the job of the user and the hands-on worker.

Parameter design

Once the bare concept has been acknowledged, and before it can be produced or delivered, two characteristics need to be identified. Working from the perspective of the customer, those features of the product that are of importance have to be clearly flagged.

When designing a new travel service, for instance, we might have had a basic concept of replacing the discount 'bucket' shops of the 1980s with a more reputable, modern equivalent catering mainly for the pleasure traveller and accessed via the internet. At a simplistic level this is our system design. The business planning process goes on to define what qualities the customer is going to be looking for. These may include:

- twenty-four-hour access
- online, accessible and very specific information

- low cost
- reputation-based guarantees rather than bonded, financial ones
- lively personality.

Traditional values, such as expertise on foreign locations, a trustworthy banking pedigree, squeaky-clean, middle-class, motherly, staff, dressed to look like cabin crew, and vast quantities of printed brochures no longer feature in the plan.

The more complex the product is, the more likely it is to have many such quality characteristics. If they are not thoroughly examined, then the product or service is not likely to succeed. In the West, this has been an area in which we have been almost totally deficient. On a typical project the Japanese will spend nearly twenty times longer than their Western counterpart. Often this activity involves extensive consumer-focused market research.

Quality function deployment is one tool that can assist in this process. It was developed in Japan and has been applied extensively both there and elsewhere. Like all other product improvement techniques it can be performed in isolation; however, its successful application depends on a team-based approach working directly with customers. This is just the environment that you would expect in an organization pursuing six sigma. Quality function deployment overcomes some of the common problems plaguing the product development process, especially when trying to dig a little deeper than the system design phase.

Where experience is limited

One major consequence of the global recession in the early 1990s was the change in the skills required by large organizations. Often this meant a loss of the long-term thinkers and builders, and their progressive replacement with short-term cost-cutters. This resulted in a glut of senior executives on the job market. Nobody would dispute the depth of their knowledge, but it is not necessarily suited to the job market in a recession. As a result many of these people decide sooner rather than later to embark on a more entrepreneurial existence. Although for some this is not a 'personal calling', for a few it presents a challenge that they would not have dreamt of in their formative years.

Ken Veit, the owner of a Cartoon Corner store in Scottsdale, Arizona, USA, was typical of those who made the change successfully. His experience establishing his own small retail outlet for cartoons and cartoon-related novelties highlighted the importance of exactly this type of planning for any venture, even the high-street retailer (Veit, 1992). His main competitors were Disney and Warner Brothers. Among the quality parameters that he identified for his own business were:

- the need to provide browsing facilities
- levels of service that no competitor could provide
- the proximity of the owner to the customer, again unmatched by competitors
- customization to suit local trade.

The growth of the internet in the later 1990s, and especially of organizations such as eBay, which cater for specialist and impromptu collectors or memorabilia, meant that the original business plan needed dramatic updating through time: another example of the importance of iterative NPD.

While no one appreciated the severity of the recession at the time, or the impact of the first Gulf Crisis, many of the lessons that Veit learnt in the first two years of operation could have been predicted. They probably were not, because in Ken Veit's case he had no experience of his market. The purpose of parameter design is to overcome that gap or to reassess the accumulated wisdom that people otherwise claim.

Where the skills and interest are lacking

Unfortunately, the skills and interests that lead people into a career in engineering, or any of the manufacturing industries, are often not those that you would associate with going out and meeting the customers.

One long-established government department which became self-financing very quickly discovered this fact. Its employees were almost entirely scientists and engineers with a talent for research work. They largely did not enjoy contact with outsiders, and so avoided it. When a firm of marketing consultants was asked to go and talk to customers to find ways of expanding the revenue for the department the message was very clear. Almost all of their customers said that they would be happy to give them more work, if only they bothered to ask for it.

One of the quality characteristics of their service was a constant building of relationships and an ongoing support role. To achieve this, their product development process took two further steps. First, the organization had to restructure to allow senior managers to focus their efforts on building relationships with specific customers rather than managing detailed scientific work. Then the individuals had to be trained to help them to sell more effectively.

This situation is repeated in many different companies. Often the point of contact with customers is the sales staff. They rarely have any contact with the product developers, and so it is hardly surprising that very little time is spent on parameter design.

Tolerance design

Armed with a picture of the final product, and with the expectations of the customer clearly identified, along with the relevant factors that will affect the delivery of these, the product developer moves on to the third phase. Whether it is the same people depends on the organization. An extreme example would be the development team associated with the Apple Macintosh computer: a cross-functional, carefully selected and preciously closeted team of people tasked with a very specific job. There are now so many good examples of these teams developing products that it is difficult to see why any organization should deliberately stifle the process by introducing unnecessary barriers.

For organizations pursuing six sigma there are some very sound reasons why they should use the same team for the tolerance design phase. The tools that they have used and the experiments that they have carried out in the parameter design phase often only need to be taken to a further degree of refinement to establish all the tolerances necessary.

Tolerances are pay-offs. They are used to decide whether the relative cost of a particular enhancement is worthwhile for the improved control that it provides. For example, at this stage our travel agents may decide whether providing a call centre open twenty-four hours a day is worthwhile, or whether to cover the period from 7 a.m. to 12 midnight. In a manufacturing context, the designer will be establishing whether to upgrade (or downgrade) a component for the increased control that this may offer.

Some of the tolerances established will be relatively simple to effect because they involve a one-off decision, such as using a single supplier of a component rather than multisourcing. Others need to be monitored continuously to ensure that they remain in control. The techniques used for this process are based on SPC. Again, they can be applied in both the manufacturing and service contexts.

References

Cracraft, J. and Eldredge, N. (eds) (1979) *Phylogenetic Analysis and Palaeontology*. Guildford: Columbia University Press.

Ealey, L. A. (1988) *Quality by Design*. Dearborn, MI: ASI Press.

Veit, K. (1992) The reluctant entrepreneur. *Harvard Business Review* 70(6): 40–49.

Yabe, M. (1990) How quality went on holiday. *Asahi Shimbun Extra, Tokyo*, 5 June.

5

Quality function deployment

Why is quality function deployment necessary?

Organizations embarking on the six sigma journey do so from one of three starting points, and it is useful to be realistic with yourself and others as to which you are coming from. If nothing else, they strongly influence the internal (and external) communications strategies that you use.

- *Internal trigger, from a position of strength*: there are those organizations that are genuinely successful, perceive the marketing advantage of exceptional quality and expect to gain some relatively small-scale improvements along the way. Under pressure of market forces, this group may or may not sustain the process for more than a few months or years. The trigger to change has always been internal and does not usually involve any 'either/or' decisions. This means that the quality improvement often becomes a bolt-on rather than a replacement for existing activities. Few, if any, really successful implementations of the six sigma approach have come from this starting stock.

- *Internal trigger, from a position of weakness*: many companies exist in a kind of limbo between success and failure. They are not under great pressure to improve, often because they are no worse than their direct competitors, but equally they cannot afford complacency. The trigger to change for these companies is often a charismatic leader. This person recognizes the longer term threat, and through his or her own analysis, recognizes the need to improve through a major step-change rather than small-scale notches.

Such companies can rarely afford the indulgence of a bolt-on process, and it is through the persistence and strength of the leader that the rest of the management team become committed to the changes that are necessary. They are making genuine 'either/or' decisions.

Very many of the 'exemplary' companies, the ones that business school gurus, pollsters and journalists love to quote as paragons of improvement, longevity or radical ways of organization, are owned and managed by one person. The extreme examples have been handed down the same family for generations. These are perfect examples of the kind who embark on a process of change because of the long-term vision of one charismatic leader. They are a tough act to follow, and few listed companies or wholly owned subsidiaries will ever stand a chance of succeeding in this kind of venture. There is no reason why they should not try, but such companies have a limited chance of success, and certainly government bodies and public utilities are extremely unlikely to succeed.

■ *External trigger*: when a business is already under extreme pressure from outside, the success of the improvement process will often depend on the commitment of the senior managers. They have little or no opportunity to make decisions; they will mainly be ratifying 'do or die' strategies.

There is a lot of debate about what makes for successful change. Most consultants have their own definitive answer. In practice, the process adopted is meaningless unless the right stimuli and triggers are present.

Quality function deployment (QFD), applied properly, will affect most of the people within an organization. As with most techniques it can be applied piecemeal, and may even produce a few successes. But it will only really achieve lasting results if it is applied as a major element in a serious quality improvement process.

Why listen anyway?

One of those fundamental principles that everyone agrees on in the quality improvement field is that 'quality' can only be defined in terms of satisfied customers. Despite this, most quality improvement processes do not involve external customers for at least the first eighteen months. There are lots of reasons for this. They include the fact that quality is often seen as belonging to 'manufacturing', which does not usually have many contacts with the external customer. This is compounded by the tendency, which is particularly common in organizations larger than about 500 employees, for the people who have traditionally risen to the top in manufacturing to be more comfortable in a technical environment than with the commercial world.

Quality function deployment provides a simple tool that is highly structured, appealing to engineers, and yet requires substantial contact with real customers.

It seems almost banal to say it, but often the last people to be consulted about a product, especially in the manufacturing sector, are the customers. There is an immense resistance to discuss specifics with customers. We often hear managers talk, albeit in whispers, about the naiveté of customers, their tendency to look retrospectively, their lack of knowledge of technical limitations, their failure to appreciate the complexity of a product or the tradeoffs involved in its design: 'they want to have their cake and to eat it, too'. Yet, most 'new' products are simple enhancements of existing ones that the customers have used more extensively than any product tester.

In the service sector, many 'new' services are not only simple enhancements; in some areas many appear to be only cosmetic repackaging. When a step-change does occur, it can often be traced back to a straightforward naive question. And who are the people best suited to ask naive questions of existing products? The technically knowledgeable creators of the original, or the naive, retrospective customers?

We already listen

There are two problems with this stance. First, who does the listening? Secondly, do they really?

Assume that they do listen. Product or service improvements, whether on a small or a grand scale, depend on translation. The person doing the listening has to translate the original, often subtle, but layperson's, customer terminology into the internal (technical) language of the organization. This may then also have to be translated from sales-speak into marketing-speak into manufacturing-speak into engineering-speak into purchasing-speak into servicing-speak and back into marketing-speak and, eventually, into sales-speak!

We have all tried the party-games, and we know that translations lose something important, or gain something unwanted, at every stage. Even remarkably simple ideas become distorted and transformed. We stumbled on one example of this with a photocopier service engineer. A simple customer enquiry asking for a blue or brown toner to go into their standard black photocopier (which originally cost £800) was taken back to base and translated (in this case through only two departments) into a customer requirement for a £12 000 colour laser copier!

The key to quality improvement is to have everyone listening to the end-user of the product and to all the people up the supply chain. This sounds impossible, yet companies around the world have demonstrated that it is remarkably effective and not all that difficult to achieve.

If you feel that you have already cracked this problem, and have in place all the mechanisms necessary to put every director, every manager and every front-line worker into contact with outsiders, then fine. As one of the tenets of the quality improvement culture is that of relevant measurement, why not just confirm for yourself that this listening process really is happening?

Hazard a guess at the right percentage of time that a person should be listening. We would suggest that 5 per cent is a barely satisfactory figure, but you may decide that factory workers (who 'only' make and prepare the final product for delivery) only need half a day each year. In the UK, half a day annually is about 0.2 per cent of the working time of such a person. You may find it interesting to use timesheets, and direct or indirect interrogation, to find out how much time is actually spent in real contact with, and listening to, customers.

For a company of 400 employees, 5 per cent of their working time would represent 4800 person-days; 0.2 per cent represents 192 person-days. In a typical 400-person manufacturing organization, the combined customer contact time of the sales, marketing, quality and customer service departments, together with the time spent by directors, rarely amounts to more than 1 per cent of the total effort of the company.

The quantity of direct customer contact is not only remarkably low in most organizations, but also often very ineffectively handled. While organizations with quality management systems have usually developed formal procedures to handle customer complaints, it is only very exceptional organizations that have any positive processes to obtain and use customer input for the development of products or services.

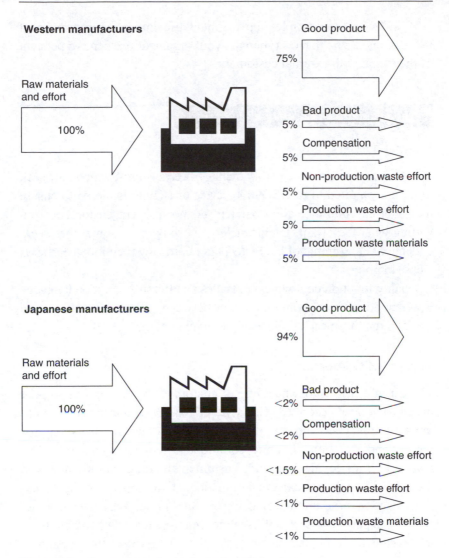

Figure 5.1 Comparison of manufacturing efficiencies

We can all recount examples of insensitivity to customers, known as 'lost opportunities'. Quality function deployment is a technique that turns these otherwise negative costs of quality into value-adding costs of prevention.

The reason why some management consultants succeed where in-house staff have failed is because they have learned to listen to customers without letting their own prejudices get in the way.

If you want one simple step to inject life into your improvement process, make sure that everybody in your organization spends 5 per cent of their time with external customers.

The problems of poor communication

Internal barriers

Many organizations with quality management systems, which they seek to have accredited to the ISO9000 series of standards, do so to ensure that internal communication barriers do not prevent customers from getting what they want. Unfortunately, these barriers sometimes seem endemic. The example of the radio-pager company mentioned earlier is a case in point.

Quality function deployment provides an effective common language to allow economical communication between different functions and across large distances within a single organization.

The 'yo-yo' syndrome

The car hire firm whose handling of customer complaints was so poor has already been described. This style of 'yo-yo' response, where problems are bounced from one department to another and between customers and suppliers, is quite common. Often there are very clear responsibilities for dealing with customer communications; however, the procedures do not exist or are inadequate to cope with the exceptions. Under these circumstances it is easy to see why the customer services department does not have the time to gather data to help them to develop the products on offer. As the customer contact is almost entirely negative, it reinforces the view that customers only look retrospectively and counter-productively.

Activity

No matter what your role is in the organization, go and spend half a shift in the 'customer service' area. Listen to the customers, check yourself becoming defensive, see if you really can resolve their problems, and do so; then see what the permanent staff have to say about your approach. If you get a very difficult customer, introduce yourself and genuinely ask them what you could do to put things right.

Example: Applause – Barts Spices, Bristol

Twelve months ago, I bought a jar of garam masala from a supermarket. I had a choice of three varieties, and not being an expert on Indian spices, I went for the cheapest. When the jar was first opened and used, the taste was awful. There had to be something wrong with it.

My partner urged me to throw it away and buy another from a 'better' brand. But I felt that the manufacturer ought to know that something was wrong. It is not that I have ever found manufacturers particularly interested, and I knew that the supermarket would replace it without question. However, for ten months, the jar sat on the desk in my study waiting for a moment when I could be bothered to scribble a note and package it up. Eventually, the day came and the jar was despatched.

A week later the postman knocked with a parcel. Inside was an extensive range of Barts Spices, an attractive booklet on how to use them and a covering letter. Not just a photocopied 'thank you' from customer services, but a personal letter from the manufacturing manager. In it he explained clearly what had gone wrong with the jar I had, what action they had taken to ensure that no others were tainted in the same way, the permanent remedial action: a new design of lid, and his personal thanks for my having brought the matter to his attention.

Will I ever pick up a different brand when there's Barts on the shelf? Will I heck: he has won a lifetime loyal customer! How? By listening, by treating the customer as a normal intelligent person and by responding intelligently. Frankly, the superb range of free products will probably go largely untouched, they were not the thing that mattered.

Example: Disaster – VW Motorworld, Kidlington

A couple of years ago, I bought an 'approved used' Volkswagen Golf car. It was the top-of-the-range automatic. Within a few weeks it had broken down, so dramatically that it was off the road for nearly four months. As it was covered by a warranty, I was given a replacement: a bottom-of-the-range diesel car. Once it had become obvious that this was not a simple fault that would go away or could be fixed in a couple of days, the service department staff became elusive; they did not return my calls or did so a day later, and they began to point the blame elsewhere: it was the insurance assessor for the extended warranty (which is really just a third party insurance policy), it was the German parts department, it was 'not something that normally happens', and so on. Eventually, in desperation, I tried ringing their service manager. He was perpetually engaged, and again did not return calls. So, I wrote to the managing director; and I wrote again; and I wrote a third time. Why? Because he could not be bothered to reply.

The customer relations department of VW in the UK was slightly better: they sent blatantly standard letters out, but did assign a customer relations agent to my case. He left after a few weeks and was not replaced. So I wrote to the customer relations department of VW in Germany. Did they reply? No!

Eventually the car was returned to me, supposedly fixed. Three weeks later it was due for a service. The dealer I took it to (hardly surprising it was not the same one) was amazed. They immediately discovered that one of the faults had not been addressed at all and another had only been worked around rather than simply replacing a small part.

Have I ever been back to VW Motorworld again? Yes, once, because there was (unbelievably) an urgent safety recall for the brake system and they were the nearest dealer. Will I ever buy a Volkswagen car again? No way.

Every management book will tell you that communication is key. Well, sadly, communication with customers comes a long way down the list for some companies.

By building the collection of data from customer feedback into the delivery process, a few exceptional organizations manage to use it as an effective product development tool, for example, the Deming Prize winner, the Joban Hawaiian Centre. Their improvement process was particularly good as it removed potential social stigma between members, and created an environment where other, more personal, improvement issues could be discussed.

This last point is important. Many 'professional' organizations, especially in the education, healthcare, legal and financial sectors, find it difficult to implement improvement processes because of a fear of provoking a defensive reaction among the staff. A letter published in *The Times Higher Education Supplement* from an esteemed academic illustrates the problem of achieving change within professional institutions:

The Director of Enterprise at Huddersfield says that quality is a 'measure of the success with which a university achieves the standard of service it has set itself by managing effectively the process of providing the service'. This is not so.

Quality in a university or of a university has little or nothing to do with management. It has to do with the professional excellence of individuals as teachers and researchers.

The use of simple, specific, customer input controlled by the staff can be very effective as a way of opening up the quality debate. Because it translates opinions into facts, QFD can be a powerful stimulus to improvement, especially in areas where dialogue does not normally exist between technical specialists, the front-line staff and customers or end-users.

Whose problem is it anyway?

Have you ever tried to complain about a product, only to find that the packaging does not give any clues as to whom you should write? Another good example of not listening to customers! So you decide to contact the chief executive on the basis that he or she will forward your reply to the right people. If it works, this route neatly overcomes one of the worst problems in handling quality improvement.

Top-down versus front-line-in

We have said earlier that, in our experience, those improvement processes that really succeed have not just the complete commitment of the most senior manager, but also his or her impassioned involvement. There has been, and will always be, a great deal of debate about what motivates people to do an exceptional job. Above all else, though, is the sincere recognition by their senior managers. Where the chief executive is involved in the quality improvement process, any complaint or problem becomes not only a challenge to the people assigned with overcoming it, but also an opportunity for those people to shine, a chance to control their own recognition. Properly motivated and managed, these people will move hitherto immovable obstacles to achieve a good outcome.

Compare this with the more common situation in which each senior manager has his own predefined activity and is not totally convinced about his or her chief executive's motives and future plans. Quality improvements that affect only their own function will be given support and resources. However, when a problem or opportunity has been identified by one function and needs to be communicated to others before it can be tackled, the incentive for people to 'rock the boat' is not only missing, but often effectively discouraged. The very people who, in the first case, would be seen as champions are perceived as outspoken critics, or at least as slightly immature in their approach.

This makes it very difficult for front-line staff, including the people who are in everyday contact with customers, to forward their ideas through the business.

Quality function deployment again provides a means by which people can advance ideas and develop the product or service, in a systematic manner, without being identified as trouble-makers.

Function versus function

How a problem is tackled depends as much on who is solving it as on how they do so. There is significant evidence that people in specific roles have many similarities in personality and outlook compared with others in the same role. These apply across different organizations, and so are not learnt within the individual business. Thus, production engineers will share similar interests and ways of reasoning, while accountants will have different approaches, but will again share these with their peers, and so on.

If a customer requirement is identified in, say, the accounts department, the approach to addressing it will probably be very different from the one that the quality department would use. Neither is necessarily right or wrong, they are just different. In the same way, a salesperson is likely to tackle problems in an equally different manner, although again they would not necessarily be right or wrong. Difficulties arise when problems affect more than one department.

Because of differences in reasoning, the ideal solution and the logic behind it that persuade a senior production manager will be different from those that persuade a financial manager or a human resources manager. As consultants, we would usually prepare different proposals for each, on the basis that they have to be persuaded individually, even though the solution may be the same. This has consequences not only in terms of how you gain commitment to your ideas, but also in how you respond to the ideas and suggestions of others (including customers); it may not be that they are wrong, it could just be that they reason differently.

In an environment that has already gone down the quality improvement route for some time, interdepartmental politics should have become an acceptable topic for discussion. If it has not, then the people proposing a solution have to be very cautious or they risk hitting internal barriers again.

Activity: Your organization's history of internal politics

If this seems overcomplicated and you do not believe it, then there is a simple exercise that can be used to demonstrate its validity.

In organizational dynamics terms, the problem is that the 'task content' is the same, but because the 'people content' is different, more than one 'task process' exists. For example, review the minutes of a year's worth of management meetings. For each assignable action, identify where the need for the action was first seen, then where it was assigned and, finally, how long it took to be completed. We can virtually guarantee that tasks identified in one place but assigned to another will take longer to achieve than those tackled in the place where they originated.

An interesting example of this came to light in the course of the Hutton Inquiry, the UK government inquiry into the circumstances around the death of a civil servant, Dr David Kelly. BBC journalist Eddie Mair revealed that the intensity of distrust even among different programme teams within the BBC meant that they would generally try to verify their colleagues' sources independently rather than trust one another. He went on to explain that his own programme team deliberately used acronyms and obscure abbreviations to prevent their peers from stealing their own 'exclusives'. How this can be beneficial to the organization as a whole is hard to see.

By providing a common language for customer-driven improvement and focusing on fact rather than opinion, QFD helps to address these problems by allowing people to share ownership of the problem and its solution. However, QFD cannot be applied effectively in isolation, as the necessary steps to a collaborative and empowered climate within the organization are vital.

Specializations

Different specialists will propose solutions that draw heavily upon their own expertise. Ask information technology people for a solution and they will propose a computer-based one. Ask the human resources department and it will involve people, whereas traditional quality departments may come up with either a change in procedures or 'improved' monitoring.

When you examine the evidence in support of each solution, it may vary in its sophistication, but each will put forward a good case. As we have said, QFD cannot be applied in isolation: it is a tool for teamwork and primarily for teams working across the functions of the organization. As it allows the effects on the customer of different technical solutions to be assessed, QFD forms an ideal answer to the conflicting claims of the different specialists.

What are we trying to get right?

Optimize the product

In the 1960s, the Western world had a scornful attitude towards products made in the East. 'Made in Japan' was synonymous with cheap goods. Although they were inexpensive, the products performed as well as customers expected them to. Initially, they were largely plastic and fabric products, but slowly the Far East began to export electronic and automotive products. At the time these products were also ascribed with the 'cheap' image, although today it is interesting to see the number of much older Nissan and Toyota cars still being driven, compared with their European contemporaries that have long since gone to the scrapyard.

Throughout this period, in the West, quality was largely controlled by inspection, in some cases by 100 per cent inspection. There is something appealing in inspecting finished, or part finished, products. Even today some Western companies still use 100 per cent inspection.

We recently visited a company making plastic bottles and caps to a tough specification for a safety critical industry. While it had invested in expensive kit, it had not done anything systematic to improve the quality of its production processes. As a result, the company still employed a woman who sat all day at a table visually checking every cap and every bottle immediately before it was shipped.

Inspection as a technique is ineffective and costly, and adds no value. Instead, it is vital to optimize the product to meet the needs of the customer precisely. As discussed in Chapters 6 and 7, this does not have to involve 100 per cent inspection. But the old adage, 'garbage in, garbage out', so popular with computer users, applies equally to production processes. If you use inferior raw materials and process them ineffectively, then you cannot help but let down your customers. The trouble is that the customers' needs are rarely expressed in terms that engineers can understand and apply.

Quality function deployment provides a valuable opportunity to introduce the customers' perception of the product and its performance into the processes that produce it. Quality function deployment does not blindly assume that we have got it right from the customers' perspective, nor does it optimize the process at the expense of the product.

Optimize the process

We shall look in much more depth at the problems of optimizing the process in the next chapter. The key, though, to the Japanese invasion of Western markets was their ability to optimize processes as well as products. Thus, not only can they deliver goods that meet the customers' expectations (in fact, often exceeding them because of the unjustified reputation), but they have done so in a very efficient manner.

Figure 5.1 showed a typical Western plant in the mid-1970s, compared with its Japanese counterpart, in terms of production efficiency. The Western philosophy of controlling quality by inspection meant that considerable waste occurred during manufacture before product inspection, in raw materials used to produce rejected product and, as inspection was often only conducted to 95 per cent levels, in returns and compensation for these returns.

In the late 1980s, not only was one textile company shipping defective product, but when customers caught them out, both the sales department and the manufacturing plant were compensating them, either in cash or by offering them the rubbish as 'seconds'!

Quality function deployment is a useful tool for linking the characteristics of products, as the customer or end-user sees them, to the technical parameters from which they are delivered. As such, it allows an organization to optimize its own processes in the confidence that they will not adversely affect the customer.

Target versus tolerance

In Chapter 6 we shall look at Genichi Taguchi's techniques for designing to targets rather than tolerances. He based his arguments on the cost of the different strategies as a 'loss to society'. This may sound rather pompous, but is probably a quirk of translation. For the word 'society' use the Western mathematical concept of 'envelope', and you are closer to his real meaning.

What Taguchi observed was that when something is designed to tolerances, statistically speaking a certain number will fall outside those tolerances and yet still be supplied to the customer. When the costs of handling these defective products are added up, we find that they are not incurred entirely by the manufacturer. Many will be paid for by the customer, and there will be others, too.

For instance, imagine a plastics firm. It buys raw materials from chemical companies. If a small proportion of its products are defective when they reach customers, then they will probably have to be replaced. The replacement cost is not just the production cost of the replacements, nor is it the lost profit on those replacements. Their lorry will have to make an additional delivery or run the next trip at a lower capacity. This lorry will emit effluents that have an impact on the environment. The additional trip adds further to the damage being done, unnecessarily. As companies begin to adopt BS7750 and audit the environmental impacts of their activities, Taguchi's 'loss function' takes on a new meaning.

When you begin to take into account all the knock-on costs in this way, they escalate rapidly as you move away from the absolute target value of the characteristic that led to the defects. This phenomenon is not just in the domain of the manufacturing company.

Consider the busy commuter train that arrives at its city centre destination five minutes late. A proportion of its passengers will be irrevocably late for important appointments or for work. If there were eight carriages in the train each carrying eighty passengers, then the delay of the train would have potentially caused a 'loss to society' amounting to some 6.67 lost person-days. As people tend to build in safety buffers in their travel, a delay of five minutes would actually only cause a few of these people, perhaps one in twenty, to be late. This means that for the five-minute delay the loss could be only one-third of a person-day. But if the delay is thirty minutes, probably more than 60 per cent of the people would have lost time. Thus, the cost to society increases not just linearly, but exponentially.

As a technique, QFD makes sure that we have understood the importance of specific customer requirements. By designing products and services to deliver them precisely, we avoid the hidden costs of quality that lead to the escalating loss function described by Taguchi.

The organizational learning curve

There is considerable evidence that simply reading linear text (like this book) is not the most effective way of absorbing information. In the same way, expecting new people joining the organization to learn quickly from traditional manuals is no longer realistic. The charts produced in the process of carrying out a QFD study provide an excellent way of presenting new staff with a summary of the knowledge that we have about a process. Toyota, which has been using the technique since 1977, has now built up an extensive library of QFD charts showing varying levels of complexity. They use these, in progressively greater detail, throughout the training of members of their staff.

The library of documents that Toyota has generated over the years has another benefit. They find that the time spent in start-up activities for new products can be dramatically reduced by reviewing the conclusions reached during the development of similar components for previous models. By employing QFD, Toyota has reduced the number of premanufacture problems by 50–60 per cent. They also do not suffer the initial surge in customer complaints that many other motor manufacturers report.

The overall saving in start-up costs that they achieved between 1977 and 1984 was estimated at over 60 per cent (Hauser and Clausing, 1988).

What is quality function deployment?

Quality function deployment is a simple tool consisting of a series of interlocking diagrams or charts, which provide a summary of the quality related information available about any product or service. The technique traces its origins back to the Kobe Shipyard in Japan, where it was first used in 1972. It was first described in 1978, and in Japan this has been the only substantial source of information on the technique (Mizuno and Akao, 1978). Published in Japanese, it remained undiscovered by Western companies for nearly two decades. Since 1967, when it was first mentioned in Kobe's internal literature, the number of known applications in Japan has grown exponentially.

Once completed and assembled, the charts used to conduct QFD look like a house, which is why it has become known as the 'house of quality'. The term translated as 'deployment' has a wide context. Whereas the typical interpretation by a Western mind would be: 'to define where something should be or happen', in Japanese it means 'to broaden the application of something'. Thus, QFD is a tool intended to broaden the responsibility for quality. It does so by identifying who is responsible for key aspects, in the customers' eyes. This is an interesting and not entirely semantic point of difference from the teachings of some Western 'quality' gurus, namely that everyone is responsible for quality, regardless of their job. It says something about the differences in culture. In Japan there is no need for everyone to be responsible for everyone else's quality, because individual responsibility is much greater. In the West, there is a need to create an environment in which everyone takes on responsibility for quality. Unfortunately, attempts to create this environment sometimes fail because a culture of 'catching one another out' can develop. In my experience this is often most marked in the 'professional' and information technology fields.

As with all quality tools, there are two aspects to QFD. First there is the technique itself, and then there is the process by which it is put into place. We looked at the importance of the organization's culture in Chapter 3. Here, we shall describe and illustrate the simple steps used to apply QFD to a particular situation. It is important to recognize, though, that QFD will do very little if it is merely applied by an individual. Quality function deployment is a team tool and needs to be applied as such.

The easiest way to describe QFD is to build it up from scratch. The collated QFD diagrams are illustrated in Figure 5.2. The process is

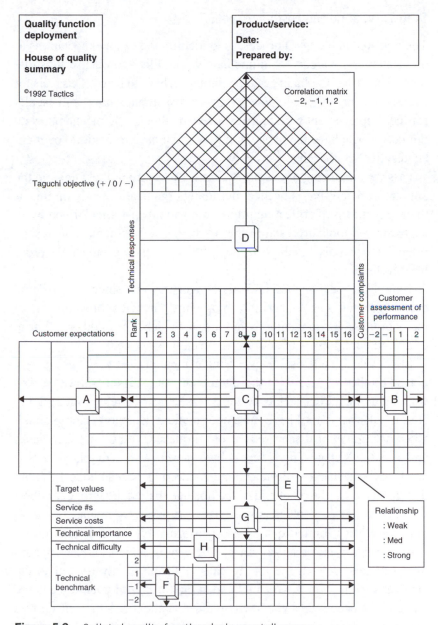

Figure 5.2 Collated quality function deployment diagrams

described in terms of the steps that lead to the final chart being assembled. Appendix 5 contains a blank chart that can be copied. You can make bigger versions to assemble into a larger format 'house of quality'.

Step 1: What do the customers want?

The first step in the QFD process is to identify the exact expectations of the customers. We can see immediately why this approach brings distinct advantages. To obtain customer input, we have to listen. This calls for a reassessment of the listening devices that we already have, and usually for their replacement with more effective methods. The mechanisms do not necessarily have to change, but they often have to begin to be managed actively rather than passively.

Passive monitoring of customer comments is rarely likely to yield the amount of detail that is needed, but many organizations rely on this as their only means of collecting information about customers. Even when sophisticated tools are used in the marketing department, their results often fail to percolate to the other departments, and especially to the product developers.

Among the tools that fall into this category are customer service desks that only handle sales or specific complaints. The telephone numbers of many of these are conveniently hidden on the label or in the small print of a document. Levels of activity are usually piecemeal.

In the 1980s, a television advertising campaign run by a lager brewery played on the lack of use of these telephone numbers. It showed a man walking along a high-tech corridor when he heard a telephone ringing. He rushed around until he found a door marked 'Customer Complaints'. Upon opening it, he discovered a room furnished with old wooden desks and a captain's chair. There were cobwebs everywhere. On the desk was an old-style mechanical dial telephone. Answering the telephone, he found that the person at the other end of the line had dialled the wrong number!

Other response mechanisms are likewise passively managed, and as such yield very limited quantities of information. Most hotels have a customer satisfaction survey form. The way in which these forms are managed ranges from the totally passive to the highly proactive. On arriving at some hotels, you are handed the form with your key and the receptionist encourages you to complete it. Some forms have incentives to complete them, such as free weekend stays. At other hotels the form is to be found hidden between a crumpled bit of hotel stationery and an old magazine in the top drawer of the desk in the bedroom. Completion rates vary tremendously. Only rarely will you be approached personally by a manager or member of staff and asked about your stay. If we are serious about obtaining customer feedback, then we can do so.

In the past few years there has been a lot of criticism of airport terminal areas. Holiday makers awaiting the departure of their flight have complained bitterly about the lack of comfort and facilities. As the airport operators began to see the (largely financial) benefit of providing extensive shopping and better quality catering, franchise operators have moved in to enhance these environments. Even so, you will have trouble finding a suggestion box.

We checked the opportunity for customers to complain at three major British airports. Only one had suggestion boxes. Supplies of comment cards had run out and there was no facility for writing on the card. The boxes were mounted in obscure places and well out of sight (and mind) of most passengers.

At the other two airports boxes were not provided; instead, we were told that passengers could always speak to a member of staff, particularly at the customer information desks. But many people find complaining, especially in person, emotionally very difficult, so it is unlikely that this approach will yield many responses. In any case, only one of the information desks had any kind of procedure for recording and subsequently analysing customer feedback, and two were unmanned.

Compare this with one European airport, where multimedia comment points enable passengers to write their concerns on a card, type them onto a screen or record them by voice.

Gathering analysable customer feedback is so rare that the places where it is solicited are much easier to recall. As with all the tools and techniques of quality, it is not what you do, but the way that you do it.

Most of us have heard of the survey conducted in the USA in the late 1970s that demonstrated that only about one in ten dissatisfied customers bother to complain. It is hardly surprising that such passive mechanisms do not provide us with enough information.

Among the more successful attempts at gathering genuine customer feedback, we should include the growth in the use of 0800 freephone numbers in the 1990s. One petrol company which decided to display its 0800 number in very large signs around its forecourts experienced a phenomenal growth in the number of people calling in. They were surprised at how consistent the causes of the complaints were: most were related to a sales promotion that they were running. It used tokens, and many people felt that they were being short-changed.

During the 1990s, 0800 numbers gave way to 0845 ones: no longer were customers given a free call, instead they had to pay the local call rate. All was fine until British Telecom finally lost its grip on the telephone system and many alternative providers emerged. Today, to people on residential telephone payment packages, 0845 numbers are among the most expensive. Do we really want to create yet another barrier to our customers?

There is still a psychological barrier to overcome if you are going to use a telephone number, whether it is free or not. For this reason many other approaches can be used.

The Nationwide Anglia Building Society introduced focus groups as a way of collecting data from its customers. Customers were encouraged to participate in Saturday morning group discussions, conducted by a trained facilitator. The results from these groups were fed back to both branch and senior management so that immediate practices and policy decisions could be reviewed.

Many companies find that one of the powerful triggers to their quality improvement process is a survey, either of customers or of their own staff. The external survey can be also used as a starting point for QFD.

Organizations vary in how they choose to conduct a survey. A government agency, the Transport Research Laboratory, chose to use external consultants. The reasons were three-fold. It was felt that consultants would act in a more objective fashion, being less inclined to become defensive under criticism. This, in turn, was expected to make outsiders more likely to respond evenly, whereas most staff would resist being critical directly to the person from the organization. The consultants' objectivity also meant that their report and recommendations would have greater credibility with the organization's sponsor, in this case the Department of Transport. Their findings were also more credible to the scientists in the laboratory, because they were perceived as experts in a distinct field.

Surveys, whether they are conducted by the company's own staff or by external consultants, can be very specific or too general. We have all experienced the lengthy questionnaire sent by a company on the pretext of getting to know its customers, which actually only consists of lifestyle questions about the consumer. This has little to do with our expectations of the product or service being developed, and a lot to do with creating a saleable database! Worse still is when these have been distributed using a firm's name on the assumption that customers share an affinity with that business and will therefore respond, whereas enquiries from a market research company would end up in the bin.

Market research can be a highly scientific process, and its reputable practitioners undoubtedly provide a depth of understanding that many in-house activities cannot. The question is really whether we want people outside our business to know more about our customers than we do, or whether less precise but more direct contact would be better.

Activity: Review your own organization's customer listening devices

Ring around your organization. Find out who collects information from customers and what they do with it.

Once we have gathered the data what do we do with them? Quality function deployment calls for a straightforward list of tangible customer expectations, such as 'easy to contact', 'keeps in touch', 'brings credibility', 'doesn't patronize' and so on. These are presented in a list forming part of the 'house of quality'. The list is arranged so that similar topics appear adjacently. This can be done in many ways. One approach that works well with groups who have brainstormed the list is to transfer the items from their flipcharts onto a fishbone diagram (sometimes known as an 'Ishikawa' or a 'cause-and-effect' diagram). There is a misconception among some people that the limbs of a fishbone diagram have to be the same regardless of its application. This is not the case and can inhibit groups rather than encouraging them; the team can choose its own themes on the basis of what the list contains (Wilson, 2000).

Another simple technique that can be very useful at this stage is the 'how–how' or 'why–why' questioning, used by many quality circles since the early 1960s. The golden rule of networking is that if you want to find out about something, you need only telephone three or four people and one of them will have all the information you need. These two techniques have a similar foundation: by asking 'how?' or 'why?' more than two or three times you will identify the definitive cause.

For example, what do customers look for in a public library? Questions to ask include: 'Why is a good catalogue important?', 'Why is an orderly collection of books important?' and 'Why are subject experts on the staff important?' After only a couple of iterations, these questions should produce the common cause: easy research. The QFD diagram would show the three original factors categorized under 'easy research'.

Such techniques are very simple to apply, but they can sound as though they are overcomplicating a simple task. However, in the more comprehensive application of QFD they are too simple.

Where QFD is applied to a complex process, such as the design of a new drug, the development of a sophisticated computer system or the redesign of a major transport system such as a regional railway network, the number of factors to be balanced can run into many thousands. At this point the human brain becomes overloaded with juggling complex details. To introduce order to such a chaotic system calls for more sophisticated tools.

One such approach is to use the family of statistical methods known as multivariate analyses (MVAs). We look at some of these in more detail in the next section, but for now we will describe only one.

Cluster analysis is one of the simpler MVAs. Like most of these techniques, it was originally developed to help taxonomists to determine the ancestry of plants and animals. Using a number of characteristics for each individual 'customer requirement', a computer arranges the requirements into groups that share the same characteristics. The nature of the groups will depend on the type of information that is fed in. This does not normally matter when working with inanimate objects, such as products or services. In fact, it can be enlightening when the computer finds a connection that you had not thought of. Hal MacFie and John Deane of the Meat Research Institute, near Bristol, UK, used cluster analysis, among other techniques, to look at the conflicting customer requirements for a wide variety of products, including cooked meats and aviation fuel (Mottram *et al.*, 1982). For example, they found that of all the tests required by different authorities for each batch of fuel, there were far fewer underlying parameters being assessed. For the company producing the fuel, this meant that whereas before they had to optimize over 40 different chemical properties, now they only had to focus on half-a-dozen.

Whereas earlier applications had to use vast mainframe computing resources, which put the technique out of reach of all but the largest corporations, today even a notebook PC can be used.

Once established, the customer requirements and their groupings are recorded on area 'A' of the house of quality.

Step 2: What are the customers' priorities?

Whether you are responsible for products or services, you will be aware of the differences in priorities that various customers place on the characteristics of the product. For one customer, a half of one percentage point extra interest will be sufficient to transfer his or her savings to another building society, while for another a courteous smile and recognition by name would be worth several points' difference.

If we are going to reassess the product or service seriously, using the results of the QFD process, then there are bound to be some tradeoffs. In practice, if you take a group of customers and let them rank the features they will come up with different lists to those of the company employees. This is a well-known phenomenon in sales training. Such courses constantly emphasize the need to understand the benefits of what you are selling as seen from the customer's perspective.

Obtaining a list of such rankings is not as easy as it sounds. For most teams applying QFD, the easiest way is to develop the list with one group of customers, and then approach a second set of customers to rank this list in order of importance. Although there are a few academic complaints about the rigour of doing it this way, practical constraints will almost always prevail.

Again, for more complex examples, MVAs can be used. Principal components analysis, for example, can be used not only to provide clustering, but also to rank the importance of each underlying component.

Step 3: What are the customers' perceptions?

The decision to set out to achieve six sigma levels of performance is based on the benefits not only to the customer, but also to the company, minimizing the cost to society in Taguchi's terms. In the short term, practical decisions about priorities for improvement need to be taken. We have already established one of the criteria in determining these; namely, the customer's rank importance of the individual factors. However, just as important in this priority setting are the customers' perceptions of where we stand in relation to our competitors.

It seems logical that improvement activities should focus on the factors that are of importance to the customers, and for which we do not perform well against our competitors. Yet, if you look at most quality improvement processes within organizations, the majority of efforts are relatively unfocused or, worse, they focus on small-scale internal issues that are within the control of a particular manager. We often hear senior people justify this failure to do anything to help the customer by talking in terms of 'starting small', learning the ropes before going 'companywide', not rocking the boat, and so on. Then, a year or so later, they put their improvement process onto the back-burner because they need to press on with more 'customer-focused' activities.

Activity

Spend a few minutes with one of the longer serving (and possibly more cynical) employees. Draw up a brief history of the company's change initiatives. What were their main characteristics? Who was involved? What worked well? What did not? Why did they end? Where are they now? How customer led were they? Did they take into account competitor knowledge? To what extent did they cross departments and functions?

There is no room for this attitude when pursuing six sigma. Everything has to be driven towards improving performance in the eyes of the customer.

How we achieve this depends on the service or product that we offer, and especially on the customer's relationship with it. Whichever route we use, the relative position of our offering is rated against the customer's, using part B of the house of quality. This assessment takes two forms: unsolicited customer comments (in practice usually complaints) and solicited feedback. The nature of the solicited feedback depends on the type of product or service, as explained below.

Customer complaints

First, statistics are collected from the customer complaints procedures. These are recorded in the first column of form B.

Think carefully before embarking on a quick call to the customer complaints department or the quality manager. A 'complaint' may not appear as such. For example, the accounts department should have a better idea of those sales accounts with a reducing turnover, or even lost ones. They will also be able to identify late and argumentative payers. While they may be belligerent, these people may also have their reasons for delaying payment.

It is useful to review the business as a whole at this stage, looking at the various ways in which complaints are received. In organizations with over 250 employees, there are usually very many more routes for complaints to be received than at first appear. This is particularly so if you start to include, as you should, the customers of your processes as well as the recipients of your product or service. For example, if you produce universal widgets, the contract drivers who transport them are customers of your processes, and just as crucial to the end-user as the product itself. How do you hear complaints from the drivers and their company? And what about other service providers? The days have passed when a company could contract out problematic services and then knobble the sub-contractor, although there are still quite a few exceptions.

One cleaning company was strongly criticized by its customer for poor attention to detail which was causing contamination of its products in certain areas. When looked at more closely, the problem was due to a combination of several factors, including a decision to reduce the amount of overhead heating. This not only made conditions difficult for the cleaners, but also reduced the flow of air through the plant that would otherwise have removed contaminants. The cleaning company management heard the grumbling of their cleaners, but did not connect the complaints with the cause, and anyway it had no channel of communication with the customer except through the shift supervisor, who was not regarded as senior enough by the client.

In another cleaning scenario, the cleaner in a small remote office was the wife of one of the junior managers who worked there. She pointed out to him that she could do a better job if she was allowed to come in at a different time. He asked his superior (it was a fairly disciplined environment) whether she could change her hours. When this information reached her managers, they told her off for 'admitting' that she was not doing a proper job.

In your review of the complaints procedures, also look out for examples of conflicting objectives. These can drastically reduce the level of overt complaints. For example, customer help-desks often record the time taken to resolve a customer's query. Two common problems arise with the statistics from these desks. First, complaints are usually tallied according to those resolved within thirty minutes, half a day, by the end of the day, and those carried over to the next day, or similar periods. This means that once a query has rolled over to the next day there is no further incentive to resolve it. So the priority for the team working on the second day is to tackle new queries and leave the rolled-over ones until their own work-load is cleared. As they are always short of staff, this almost inevitably means that these queries remain unresolved for very long periods.

The second problem arises with calls that involve a callback. The initial query is made, a tentative solution is offered and the customer goes off to try it. The assumption with most help-desks is that the customer will call again if the solution does not work. This means that the call can be removed from the monitoring system as resolved, thereby removing any incentive for the staff to pursue the customer to check that it did work. When the customer does call again, the system creates a new job and the severity of the problem itself is lost.

Broad experience products and services

When customers have experience not only of your own products or services but also of those of your competitors, then the assessment of performance can be based on direct comparison. There are many variables. For example, car manufacturers often carry out competitive comparisons involving their customers. However, as most car drivers retain the same vehicle for three years, their comparisons are likely to be well out of date and certainly not based on contemporary models. By contrast, hotel guests, especially business people, will probably have relevant, recent experiences of competitors' establishments, even if the hotels belong to the same chain.

In this case, the customers are presented with a list of customer expectations. They are asked to rate the performance of a given product against those of others. The assessment consists of a simple four-part scale. This deliberately eliminates 'don't know' types of response, but allows customers to give 'same as' responses. There is a comprehensive range of statistical techniques for the description and comparison of these scales (Kendall, 1970).

Limited experience products and services

When the customer is very unlikely to have sufficient experience of competitive products to make realistic comparisons, the question posed needs to be changed. We now ask them to rate our product for each of the expectations in part A, using a similar four-part scale, against their own perception of what should be delivered.

This can be problematic. As customers have little experience of alternatives, they may be prepared to award high marks when technical performance is actually much lower. Some researchers attempt to address this by exposing customers to some alternatives, often in the form of written descriptions or video material, before asking them to carry out the assessment. Unfortunately, this process can desensitize the customer and produce very confusing results.

One approach that can produce usable results is to present the customer with four simple descriptions, without explaining which one matches your own product.

For example, 'What proportion of first class letters should arrive the next day?'

- all first class posted letters should be delivered the next day
- nine out of ten
- eight out of ten
- seven out of ten.

There are still problems with this approach, but the results are useful.

Step 4: How good are the technical responses?

So, now we have a more tangible appreciation of how customers look at your products or services, in terms of both the features offered and their delivery. The next question concerns how your organization perceives what it delivers. In technical terms, what do you offer?

We are not yet asking whether these are the right things to do, but instead how the product or service is delivered. For complex technical applications this list could be extremely long. Some Japanese applications of QFD use between fifty and 100 technical specifications. The way in which the list is developed depends on the complexity of the product or service, and on whether you are developing a new product or an existing one.

As an illustration, the following are typical of the sorts of technical response items:

- car seat position
- switch knob type
- door seals
- interior noise.

Notice that these are not the actual specifications, but the areas in which specifications exist. The specifications themselves can be listed later. If you are developing a new product and have only a few precedents, then the QFD team will take each customer expectation and identify a few different technical features that are responsible for producing the desired result. If you are concerned with enhancing an existing product, then the team will deliberately not work this way, but instead gather information about the controls and influencing factors that are currently used.

Where the list appears to be growing beyond a manageable level, it may be necessary to split it in some way; this is a reasonable idea for this part of the house of quality. Often the preparation of technical specifications involves a number of distinct departments, in which case these departmental boundaries can be used to reduce the number of categories on a single chart. One complex example developed by Toyota used 16 subsets of this chart to look at one product feature, and each was created by the product team responsible for its development.

Gathering this type of information is usually much easier than customer data that we have been considering until now. There is often an abundance of technical literature within organizations, and it is sometimes tempting to avoid overburdening the team with data by asking only for the details that relate to the established customer expectations. The problems in doing this should be fairly obvious. The QFD approach is almost as useful for the unnecessary activities that it highlights as it is for the missing ones. It is not uncommon for much of the information, especially technical specifications, to be redundant, as shown in the next stage.

Between the list of the technical responses and their correlation matrix (see step 6) is a row labelled Taguchi objective. We shall look at this in more detail in Chapter 6; however, at this stage we should be able to identify what is the most desirable outcome for each technical response.

There will be three types of outcome. Some factors will be best when they are minimized. For example, the technical response of 'interior noise' in a car design would probably be best minimized, in which case we would mark this column with a minus ($-$) sign. Other factors, such as brake efficiency, should be maximized. These are recorded with a plus ($+$) sign.

The third type of factor will have an optimum value somewhere between the two. We usually record these with a letter '0'. Steering wheel resistance would be a case where too sloppy a wheel would be dangerous and too stiff a wheel would be too difficult for some drivers to turn.

As with all of the elements of a QFD study there is a danger in glossing over the details. For many technical responses the desirable condition will appear at first sight to be obvious. After all, who would want a noisy car interior? Well, of course, damping out all noise is technically possible, but doing so would prevent an emergency vehicle's siren from being heard, and would ruin the acoustics for increasingly sophisticated in-car entertainment systems. One solution is not to remove all noise through acoustic damping, but instead to fill the car with noise, albeit white noise, which would still allow sirens to be heard. So here the technical response, which might on first pass have been marked as a straightforward minus, is actually an '0' for optimize.

Step 5: How do the technical responses relate to customers' expectations (relationship matrix)?

We begin our analysis of this information by looking to see how internal, technical responses relate to the customers' expectations. For example, you may specify that an automotive diesel fuel should contain less than so many parts per million of inorganic particulate matter. This has a direct relationship to the customer expectation that the fuel should not clog fuel filters. This relationship is recorded in the rectangular area of part C of the house of quality. The symbol used varies: some people use a triangle to show a weak relationship, a circle for a moderate relationship and a double circle for strong relationships. There are advantages in using symbols to illustrate relationships when you are going to use the diagram as a discussion tool with other people.

We prefer to use objects that have an increasing density or size according to the strength of relationship, so weak may be given a dot, medium a large dot and strong may block out the box entirely. Alternatively, you can use a scale from zero, for no relationship, upwards. Whichever system you adopt, record it clearly on the house of quality chart, as confusion is often created through the use of unclear symbols.

If you are unsure about the relationship (or lack of it), it is well worth having a symbol for 'unknown' (a question mark will do). These relationships can become priorities for the optimization experiments described in Chapter 6.

Reviewing every possible combination between the customer expectations and the technical responses can be a very tedious process, but it can also give an immediate set of priorities for the quality improvement team. Any row that is free of symbols, or has only zeros and or question marks in it, is effectively an uncontrolled customer expectation. Either we have overlooked some technical responses from our list that we should review, or the gaps should be plugged as soon as possible.

Similarly, any column that is free of symbols, or has zeros, is potentially (although not necessarily) a redundant control mechanism or specification. If it has question marks then these become candidates for the experimental work described in Chapter 6.

Step 6: If the technical response is changed, what will be the effect?

The second step in analysing the data that have been collected involves looking at the technical responses and identifying relationships between them. For instance, in an investment bank the factors used to ensure high returns on an account may include immediate monitoring of foreign markets and shift patterns of investment staff. These two factors are related, in that by having the right shift pattern the bank can monitor foreign markets more effectively.

For an automotive parts manufacturer, the type of chemical additive used and the volume of the main component in an injection moulding operation may be related. If large volumes of raw material are consumed, then it may be necessary to use a different additive to the one best suited for smaller mouldings.

A few practitioners have suggested that this step is voluntary. In our experience it can produce some of the most significant benefits by removing duplicated effort from the production process, especially where each technical parameter is being tested.

Not only does this step produce savings in reduced effort, but it is vital in order to optimize customer expectations. The diamond-shaped matrix that forms the roof of the house of quality is used to record correlations between the technical responses. Although graphics and symbols can be used, the number two (2) can be used to indicate a strong positive correlation, say one with 99% significance, and a number one (1) to indicate a strong positive correlation with a lower degree of significance, say 95%. The numbers minus two (−2) and minus one (−1) can then be used to indicate their equivalent strong negative correlations. A perfectly acceptable alternative would be to record the actual correlation coefficients, or simply (+), 0 and (−) where this is the limit of our understanding.

One important distinction between organizations using QFD as a basic quality improvement tool and those that are pursuing six sigma is that the latter will probably insist on such correlations being statistically significant and with these levels of confidence.

Armed with these correlations it is possible to predict immediately the impact of changing one technical response on the other technical responses. For example, if you know that there is customer dissatisfaction because of a particular factor, then you can see that this is going to be improved by varying a number of technical responses. Using this last matrix you can see whether the changes will have any unexpected effects on other customer expectations.

There is an understandable reluctance to become involved in this kind of relationship analysis. Sadly, all too often it highlights how unaware we really are about our processes. There is often a lot of collective wisdom in organizations, but it is equally often out of date or pure myth.

Step 7: What are the specifications for each technical response?

You have probably accumulated most of the information available for the house of quality, part E. The data recorded here are the specifications for each of the technical responses already documented in part D.

Before the chart is completed, it is usually expanded to allow for hitherto undefined technical responses that are needed to provide checks where none existed before. These will be situations where a customer expectation had no symbols against it in part C.

For each technical response the specifications need to be identified. Again, priorities for improvement become apparent very quickly. If a technical response is supposed to ensure that a given customer expectation is being met, and yet no specification exists for it, then it is impossible for this to be effective. Remember that achieving six sigma is never going to be possible if significant details like this are not addressed.

Once the team carrying out the QFD analysis has established these technical specifications, it is quite common for them to begin to check how well they are actually monitored. This can be a time-consuming exercise, but it is a very useful activity for members of the group who have not perhaps been involved to any degree. As it means talking to most of the operators in a plant, it can also be a very constructive activity for the front-line members of the team. They will arouse more interest than suspicion when they do so.

This last activity can identify some fascinating instances of wasted effort. One team working in a printed circuit board manufacturer found that there were roughly three times the number of variables being charted than were necessary. Not only did this clog up the system with paperwork, it also meant that serious problems could be missed. This is not a common situation, as most manufacturers barely apply any form of statistical process control to their processes.

If, when carrying out this step in the QFD analysis, you find that many of the technical responses are not measurable, then it is likely that they have not been defined in sufficient detail.

Step 8: How do your competitors compare technically?

Part F of the house of quality is the foundations. Without stretching the analogy, it is also the point where the exercise is most likely to be undermined.

The technical benchmark is a comparison between your products or services and those of your competitors. However, unlike the assessment made by your customers, this one is based on the technical responses. In our experience, organizations vary tremendously between one another in the extent to which they gather this information. Competitive benchmarking has nothing to do with industrial espionage!

A lot of information can be collected about the performance levels of your real competitors. This can come from market research surveys, debriefing new employees, the trade press, and so on. But organizations pursuing six sigma will not spend much time engaged in comparisons of this kind.

Today's approach to competitive benchmarking is to compare your performance not with that of your competitors in the marketplace, but with organizations using the same, or similar, processes. These process competitors work in a different market and have inherited a different set of constraints. This means that dramatically different levels of performance can become acceptable.

For example, consider distribution. If you are to achieve six sigma levels of performance in your distribution activities, then it will only happen by learning from people who do it better. One petrochemical company was satisfied with its performance. It outperformed other petrochemical companies in most respects. For example, 85 per cent of its deliveries were made on the day that was agreed with the customer. This was not necessarily the day that the customer originally wanted, nor was it necessarily a convenient time, and it meant that 15 per cent of the deliveries were not to the customer's satisfaction, but it was significantly better than the competitors. Therefore, normal competitor benchmarking would have been reassuring but unlikely to lead to improvement.

One of the problems confronting the organization was that no one had ever expected them to do better. Newspaper distribution is a different matter. Here the product has almost no value if it arrives on a different day, and most consumers expect it to arrive before 8 a.m., otherwise they cancel their order and buy one casually instead. Yet the newspaper production process cannot really be completed until after the last television news broadcast, otherwise the newspaper would no longer be competitive.

An international express delivery service, such as DHL or Federal Express, potentially has more freedom, but their direct competitors are constantly putting pressure on them to cut delivery times while sustaining high levels of accuracy. Thus, DHL delivers over 98 per cent of its customers' packages to their satisfaction.

Similarly, the petrochemical companies deliver on order, but expect their customers to prebook, in other words to anticipate their short-term sales volumes. Other industries do not have this luxury. The ambulance service in central London, for instance, responds to nearly 1000 calls per day. It is required to do so within 14 minutes of receiving the call in 95 per cent of all cases. More recently, the specification for their service has changed and they have to achieve an eight-minute target.

The 95 per cent goal does not seem too impressive when we are discussing six sigma levels of 99.997 per cent, but it is still much better than the petrochemical business.

Another example comes from the accountancy profession. Frequently called upon to make presentations to customers, the organization acknowledged that it was staid, lacked polish and did not totally engage its audience. They did not want to introduce a comedy act, but felt that they were not selling their message properly. Using a human resources consultant as facilitator, they benchmarked sales presentations with an advertising agency and a television production company. The results were acknowledged by all three to have improved their individual performance.

Most processes can be benchmarked in this manner, and the specifications for technical responses should be too. This calls for a great deal of creativity, but the temptation to contract it out is best resisted. As with all the quality tools, once the skills have been developed in-house, the real benefits come when the people who are involved in the job carry out their own assessment and implement improvements.

Benchmarks should be technically sound, even when you are looking outside your industry. Even seemingly very specific parameters can be usefully benchmarked outside your own industry. For instance, Boeing compares in-cabin aircraft noise levels with those in other forms of transport. When you compare the sound levels inside an aircraft with those in an express train or a luxury car, you begin to see things from the customer's perspective.

The technical benchmark is shown on part F of the house of quality. The scale on the form is marked 2, 1, −1 and −2 for relative performance. If appropriate, this can be changed to a more specific scale, but avoid overspecification. The key is relative performance. If one of your comparison organizations is better than you, then show yourself as a (−1); if they are very much better give yourself a (−2).

If the technical benchmark does not give you a clear picture of the priorities for improvement, then something has gone seriously wrong. By comparing your technical benchmarks with the customers' assessments of your performance (part B), you can quickly identify those features of your product that you think are performing well against your competitors and against those that your customers think perform badly. To do this, move up the column for each individual technical response from the benchmark to the correspondence matrix (part C). When you reach a relationship between the technical response and the customer expectation, move right to the area of customer assessment (part B). There are four possible significant outcomes (see Figure 5.2).

Most of these outcomes should be self-explanatory. One useful application of this cross-checking is to validate your assumptions about the relationship between technical responses and customer expectations. For example, as brewers, you may believe that head retention (a measure of the length of time the froth remains on the top of the beer) has a relationship to taste. You would then expect that if you score highly for head retention in your benchmarks, that you should also score highly for taste.

If you do not score highly for taste, then your benchmark could be inaccurate, and either the relationship between head and taste could be less marked, or the way in which taste was assessed was questionable.

Step 9: What incentives are there to change?

The technical responses are now qualified by adding some measures of their importance. The first of these is a record of the number of service calls or returns made as a result of each technical response. The number of calls may not be a measure that you can directly relate to in your organization. In some organizations, customer complaints make a satisfactory alternative, in others changes of specification or extras can be used. Occasionally this can be measured as the actual number of events against the technical target number.

Against these service calls we also record a cost element, based on the time and materials used to correct the technical response. These can be assessed in many different ways, but if no figures are immediately available a unit time and cost basis can be used. For instance, a wire works apportioned its quality department overhead across its time-sheeted activities, thereby arriving at a cost based on each product line. These costs were then broken down according to the proportion of problems caused by similar faults.

Sometimes the specification of a technical response will be based on legislative or similar requirements. Where a technical response is not negotiable in this sense, it is acknowledged in the row marked 'technical importance'. For example, if we are concerned with engine performance and identify fuel octane rating as a technical response, then we have very clear legislative requirements that dictate the absolute values of this factor. The presence of these would be acknowledged by a mark in the technical importance row.

Step 10: How tough will it be to change?

The final assessment used in the house of quality, and recorded in part H, is the technical difficulty in making changes to each technical response. This assessment is usually made on a relative basis, such as a scale of 0, 1, 2 and 3, although there are many other alternatives.

In devising strategies to deal with poor performance in certain customer expectations, use this scale of relative difficulty to select the easiest alternative, or to decide on the order in which improvements can be made to achieve maximum effect for minimum effort. Table 5.1 is a summary of the QFD steps.

Quality function deployment applied

Nested quality function deployment

From time to time the application of QFD may seem too large to tackle as a whole, and common sense dictates that it should be broken down into bite-sized chunks. In this section one special case will be reviewed.

In general, if the product or service with which you are concerned is sufficiently self-supporting, then using QFD on it as a discrete activity

Table 5.1 Summary of quality function deployment

1. What do our customers want?
2. What are the customers' priorities?
3. What are the customers' perceptions?
4. How good are the technical responses?
5. How do the technical responses relate to the customers' expectations?
6. If the technical response is changed, what will be the effect?
7. What are the specifications for each technical response?
8. How do your competitors compare technically?
9. What incentives are there to change?
10. How tough will it be to change?

will work satisfactorily. For example, on a large scale, one of the best known applications of QFD was by Toyota in the late 1970s, which involved reducing their substantial annual bill for warranty claims due to rust. In this case the application concerned a vast array of discrete causes of rusting. The Toyota team broke the problem down into specific aspects, such as body panel corrosion, door corrosion and dysfunction of components (e.g. window-winding mechanisms) due to corrosion. Eventually, the output from all of these applications was pooled so that a comprehensive model could be constructed, but this was not before many discrete improvements had been achieved.

As always, you should bear in mind the culture in which you are trying to apply QFD. No matter how specific the application, it calls for a multidisciplinary team approach. This technique is not to be used by individuals working on their own, nor is it concerned only with creating pretty charts. There is always a strong temptation to avoid the problems of conflicting viewpoints that are inevitable in a team, and trying to 'go it alone'.

Short-paths

One model of new product development that has been heavily promoted by the American Supplier Institute (ASI) involves an iterative process consisting of four stages:

1. product planning
2. part deployment
3. process planning
4. production planning.

(Readers should note that the American terminology used by the ASI is not the same as that in common use in the UK.)

If you subscribe to this model, then it is possible to consider using QFD in four different, again iterative, loops corresponding to the four stages above:

■ *product planning*: uses steps 1, 2, 3 and 4, producing three types of technical response (component parameters, process parameters and production parameters

- *part deployment*: uses the component parameters, taking them through steps 4, 7 and 9
- *process planning*: uses the process parameters, taking them through steps 4, 7 and 9
- *production planning*: uses the production parameters, taking them through steps 4, 7 and 9.

It should be fairly obvious why we have qualms about using this approach. In our experience, many of the benefits of QFD stem from the removal of non-added value activities in step 5, and from the stimulus to improve provided by benchmarking at step 8. As this approach misses both these steps, it loses much of the potential opportunity for improvement.

References

Hauser, J. R. and Clausing, D. (1988) The house of quality. *Harvard Business Review* (May–June): 63–73.

Kendall, M. (1970) *Rank Correlation Methods*. London: Griffin.

Mizuno, S. and Akao, Y. (eds) (1978) *Quality Function Deployment – An Approach to Company-Wide Quality Control*. Tokyo: JUSE.

Mottram, D. S., Edwards, R. A. and Macfie, H. J. H. (1982) A comparison of the flavour volatiles from cooked beef and pork meat systems. *Journal of the Science of Food and Agriculture* 33: 934–944.

Siegel, S. (1956) *Nonparametric Statistics for the Behavioural Sciences*. London: McGraw-Hill.

Wilson, G. B. (2000) *Problem Solving*. London: Kogan Page.

6

Taguchi's techniques

Introduction

Once you have established the key parameters for the product design to meet your customers' expectations, then it is time to set up the production process.

One approach to achieve this was developed by the Japanese telecommunications engineer Genichi Taguchi, in the aftermath to the Second World War. It has been used extensively in Japan ever since. In the USA the American Supplier Institute introduced it commercially in the 1970s and 1980s.

In practice, Taguchi's techniques involve the use of a set of tables that can be applied by almost anyone; very little training is needed. These tables represent the statistical tool known as fully saturated partial factorial designs. Their popularity lies in the ease with which they can be used and interpreted. Once you have mastered a few simple rules, you can apply different designs to quite complex (indeed very complex) problems.

One of the first of these rules is that you should never try to work on your own when tackling a particular problem. Taguchi developed these tools for use in a Japanese culture in which team-based decision making is particularly important and they depend on the breadth of ideas and imagination that a team brings. Frequently, I have come across individual engineers who have tried to apply them on their own and have floundered. Sadly, they can waste a lot of time and effort, only to discover that they have missed a crucial factor. With Taguchi's approach, not only do they know that they have goofed, but they even know by how much; but more on that later.

Taguchi packaged his tables, and his rules for problem solving, with a number of other conceptual tools, including the 'loss function'. For anyone interested in understanding the full implications of this and why it was so important in the earlier stages of total quality in the West, Taguchi's own books have been translated into English (Taguchi, 1968).

In the pursuit of six sigma, the concepts on which these are based should have already been well absorbed within the organization as part of the six sigma culture described in Chapter 3. This chapter focuses only on the use of the tools themselves. Before doing so, it is useful to appreciate why designed experiments are so important. Such tools have never been widely available within industry. Only the larger organizations have ever had access to the necessary skills, and as these businesses represent such a small proportion of the gross national product (in the UK at least), it is not an idle claim to suggest that only a few industrial processes are running at anything approaching their potential effectiveness.

We often hear that the manufacturing sector in Britain is close to extinction. While that may be an exaggeration, experimental design is virtually unknown outside this sector. When you realize the power of the techniques, it is almost beyond belief that they are never used in the public sector (especially in railway management and the health service), in financial services, catering and hospitality, or marketing. Organizations of all kinds are desperate to reduce costs or increase profitability yet they do not bother using such simple tools to transform themselves.

I wanted to make sure that this book would give a comprehensive account of how to achieve six sigma. This has meant describing exceptions and more complex problems, as well as the basics. This is particularly the case for Taguchi's techniques. If you are new to the idea of experimental design, read the next few short sections, up to 'Analysis of variance (ANOVA)'. Stop there and catch up on the statistics later.

The case of the young scientist

Let us begin by looking at the case of a young scientist working on a pilot plant in a chemical factory. Assigned the task of optimizing the

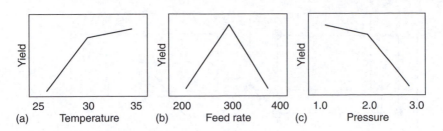

Figure 6.1   Experimental results

plant for three variables, temperature (T), pressure (P) and raw material feed rate (V), he chose to tackle the easiest variable to control first: temperature.

Measuring the batch process yield, Y, he began at 25°C, the lowest end of the feasible range, with the pressure and feed rates set at their normal operating levels.

For the next batch, the temperature was increased to the mid-range point while the other two factors were kept constant, and the yield was measured.

A third batch was prepared, again with pressure and feed rate held constant, but this time with the temperature at the upper end of the operating range. The results are presented in Figure 6.1(a).

The yield increased with temperature, but the small gains above the mid-range level were not considered beneficial, and so the scientist chose to retain this value.

Taking feed rate as the next variable to study, the scientist fixed the temperature at the optimum he had established, and set pressure at the normal operating value. Slowing the feed rate to the lowest level possible, the yield was considerably lower than normal. He repeated the experiments, but with feed rate at the norm, to obtain a measure of this yield. The sixth trial was at a higher than normal level, and the yield was low again. Figure 6.1(b) shows this relationship.

Setting the feed rate at the normal mid-level and the temperature as before, the young scientist began his seventh trial with the pressure again at the usual level. The yield was measured and the pressure gradually increased. At 1.2 atmospheres (atm) the yield had dropped slightly; when the pressure was increased to 1.4 atm it was considerably lower still. This relationship is shown in Figure 6.1(c).

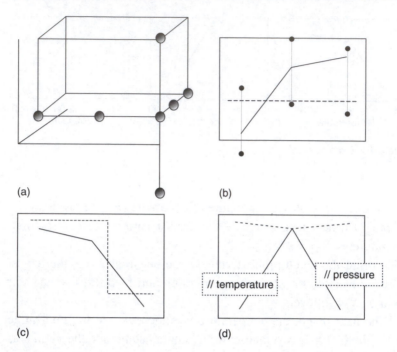

Figure 6.2 Problems encountered

The young scientist was justifiably satisfied. In only nine trials, he had shown that all the established settings were correct, and he reported this information to the research director. The research director was reassured, but puzzled as he was used to the pilot plant returning odd results. A weekend chat with a friend from a drug company confirmed his worst fears.

Before destroying the reputation of the young scientist involved, it should be acknowledged that what he did was probably far more systematic than many of his predecessors' attempts.

Let us look at a few of the problems that may have been encountered. First, we should examine how many options this scientist checked. The three-dimensional representation given in Figure 6.2(a) shows that his experiments only examined half of the possible optimum combinations.

He only repeated one of his combinations, but that gave him an impression of the variability in his experiments. The extent of this variation may mean that the apparent effect of increasing temperature is deceptive (Figure 6.2b).

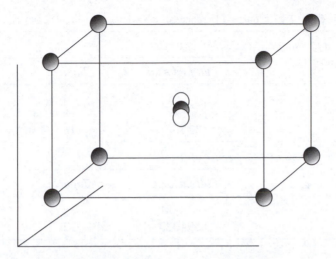

Figure 6.3 Full factorial experiment

The 1.0 and 1.2 atm experiments were conducted on the first day, whereas the 1.4 atm experiment took place on the following day. The young scientist did not know whether anything happened overnight that might explain the drop in yield in the last trial (Figure 6.2c).

The feed rate variation may have been as the scientist described; alternatively, at lower temperatures the yield may be related to temperature, whereas at higher rates it may be related to pressure (Figure 6.2d).

Properly designed experiments can overcome all of these problems, and others. The traditional Western approach would be a full factorial experiment. In this case, the approach would have called for the array of 11 points shown in Figure 6.3, two more points than the original experiments used.

By comparison, the Taguchi designs would have called for eight trials, although more trials would be a bonus. The savings in effort escalate as you deal with more factors (Table 6.1). When dealing with problems at more than two levels (only really necessary when factors have a non-linear effect), the savings are even more dramatic.

Traditional experiments versus Taguchi's techniques

It is useful to remember that Taguchi's techniques come from an industrial background rather than an academic one. Similar tools were devised

Table 6.1 The resource differences of traditional and Taguchi experiments

No. of factors at two levels	No. of trials	
	Full factorial	Taguchi
3	8	4
7	128	8
15	32 768	16
No. of factors at three levels	No. of trials	
	Full factorial	Taguchi
4	81	9
13	1 594 323	27

slightly earlier at the Rothamsted Agricultural Research Station in the UK as part of the effort to boost production of crops. Unfortunately, these tools remained in the realm of the statistician, outside the grasp of most commercial applications.

Taguchi was concerned with designing quality into something rather than fixing poor quality, although they can be used for both. The Taguchi techniques are therefore concerned with removing causes of variation rather than reducing their effect, and with minimizing variation to optimize on a target rather than fitting within tolerances.

Unlike traditional methods, the use of Taguchi's techniques does not call for the statistician's language of hypothesis and null hypothesis. Similarly, the techniques do not assume special distributions, such as 'normality'. This is because they are intended as practical tools for the real world, and not as aids to primary research. Finally, they generate reproducible results; indeed, the reproduction process is very much a part of their use.

When they first appeared in English, Taguchi's techniques triggered a reaction from statisticians. Fortunately, this has now been widely dismissed and the tools can be applied with little fear of academic challenge. Above all else, it is important to recognize that they are practical tools that provide a pragmatic self-check if applied properly. In this case they can certainly do no more harm than doing nothing, and the benefits can be enormous.

For a more detailed comparison of traditional experimental design techniques and those of Taguchi, there are management reviews of the subject that date back to their introduction to the West (e.g. Wilson *et al.*, 1990).

Points to consider when designing an experiment

Resources

The first step in designing an experiment should be a clear identification of the resources available. The people and materials, or overall budget, will limit the scope of the experiment. Without adequate resources, an over-ambitious programme is likely to fail to produce any usable results. It is often better to sacrifice answering every question for the sake of really solving a few.

The culture of improvement was discussed in Chapter 3. It should be obvious now that any tool or technique can be applied mechanically by an individual. As with all of these approaches, the real results only happen when they are applied by teams in a culture of continuous improvement, and where managers encourage rather than control. The people who will carry out the experimental work should also be involved in the design process.

Goals

The exact purpose of the experiment needs to be clearly stated. The more precise the definition is, the easier it will be to design, perform and report the results. We will see later that the type of analysis used also differs according to the nature of the problem.

Factors

The next step is to identify all the factors that will be analysed. With a group of people involved in the design, it is likely that individuals will identify different sets of factors for study. The Taguchi methodology provides a very easy way of handling these differing interests, and at minimal extra cost. By incorporating results from quality function deployment, many potentially significant factors will already have been established or dismissed.

Levels

Experiments are carried out using contrasting levels of each factor, such as high and low settings of an instrument, alternative suppliers or different standards of service. Once the factors have been chosen, appropriate levels should be selected. Only levels that would be within the working process should be used. If the results indicated that a level outside the normal range would produce an improvement, yet it is not a practical application, then the experiment has been a waste of resources.

The difference between the levels has a high correlation with the significance of the factor. If the levels are too widely spread, then almost any factor might appear significant. If two levels are chosen, then they should always fall within the normal range for the process. As a rule of thumb, the 5% and 95% points in the normal operating range usually work well.

Outcomes or responses

In general, more than one outcome should always be measured. It is a waste of resources if the trials are carried out and only one outcome is measured. Although it is common for a single response to be the most important, there are usually several that are of interest to the experimenter. The result of the process is called an outcome.

Most processes have many important outcomes. In a brewery, for instance, volumes of beer and carbon dioxide, colour, haze, taste and so on, are all important and can be optimized.

We shall look at ways of evaluating multiple characteristics later. This is important because the conditions which are optimal for one response variable may not be the same as those for another.

Number of repetitions

The number of data points that will be analysed for each experimental run must be determined. This is referred to as the number of repetitions, and is often determined by cost. The rule is 'the more repetitions the better', as the result should be closer, quantitatively, to the 'true' result. However, as it is possible that the product from the experiment may have to be scrapped, it is important that the real cost of repetitions is recognized.

Randomization

If possible, all the trials should be randomized. In the real world this is not always possible. Some factor levels can be very difficult to change, and so it may be necessary to randomize the trials in two or more blocks. The way in which the trials are randomized should be properly discussed by the team running the experiment.

For example, suppose that it is very expensive to change the operating temperature of a plant. Placing the temperature in column 1 of an L_8 array would mean that the first four trials would use one level and the second four the other. Randomizing the two blocks of four trials will overcome any possible bias without becoming a hindrance to the experiment.

Logistics

The final step in the design of an experiment should be to discuss the logistics of the trials.

- Who will conduct the trials?
- When will the work be carried out?
- Over what period?
- Who will gather the data?
- Who will carry out the analysis?

Orthogonal designs

As the number of factors and levels increases, so the complexity of designing and analysing such an experiment increases. Taguchi's approach allows us to set up experiments with a very large number of variables in an easy 'cookbook' fashion. Setting up these multifactored experiments is done using orthogonal arrays. These arrays were originally developed by statisticians, and are also known as fractional factorials. Taguchi prepared the tables in a more user-friendly form, so that only parts of the fractional factorial are used. To understand what is meant by an orthogonal array table, we shall examine each of the elements of a typical table. This example is called an L_8 array (Table 6.2).

Table 6.2 Orthogonal array table, L_8

Trial	Factor						
	A	B	C	D	E	F	G
	Column						
	1	2	3	4	5	6	7
1	1	1	1	1	1	1	1
2	1	1	1	2	2	2	2
3	1	2	2	1	1	2	2
4	1	2	2	2	2	1	1
5	2	1	2	1	2	1	2
6	2	1	2	2	1	2	1
7	2	2	1	1	2	2	1
8	2	2	1	2	1	1	2

Eight experimental 'trials'.
Balanced number of 1s and 2s.
Any pair of columns has only four combinations: (1,1), (1,2), (2,1), (2,2).
If the same number of these combinations occurs, then the columns are orthogonal.
In the L_8, any pair of columns is orthogonal.
The L_8 can be applied to seven or fewer factors.

First, the notation for this array is L_8 (2^7). This means that there can be up to seven factors with two levels in the design. The number eight shows that the design will require eight experimental trials to produce analysable results.

This array is the equivalent of a 2^7 factorial experiment, which would require 128 trials. The Taguchi design only requires eight trials, so it is clearly more efficient. This form of notation is identical for all the orthogonal arrays.

The orthogonal array has seven columns, each labelled with a letter. This refers to the seven factors that can be used in the design. In real problems, the proper name of the factors (temperature, pressure, etc.) would be used. Smaller numbers of factors can be used. For example, if you only want to study three factors, the smaller array (L_4) will do.

To the left of the array are the numbers of the trials. The number of trials on any particular array does not change even if you use fewer than seven factors.

In each column there are four 1s and four 2s. The 1s and 2s refer to the first and second levels of each of the factors in the design. These could be filled in with the appropriate settings for these levels for each factor.

For example, if the first level of temperature was 250°F and the second 270°F, then wherever there is a 1 in the table this could be substituted with 250°F, and similarly the 2s could be replaced with 270°F.

There is a balancing property between the columns in the table, which guarantees that the conclusions reached about the factors will be independent and uncorrelated. This is known as orthogonality. Orthogonality means that between two columns each combination of levels occurs the same number of times. Taking columns 1 and 2 in the L_8 array, the combinations (1,1), (1,2), (2,1) and (2,2) each occur twice. Any two columns in the tables will be orthogonal to one another, which is why the tables are said to be balanced.

By reading across the rows, we determine which levels of each factor should be used for one particular trial. So, the first trial in the L_8 (2^7) example calls for the lower level of all the factors. The second trial uses the lower level for factors A, B and C, and the upper level for D, E, F and G.

Although we commonly refer to lower and upper levels, this is rather loose terminology, as there is no reason why level 1 should not represent a higher physical value than level 2. For example, level 1 in this case might be 270°F and level 2 might be 250°F.

For each trial the outcome, sometimes called the response, is measured. As noted, several different response variables would normally be measured for the same trial. This is obviously more cost-effective than running separate experimental programmes for each. This may mean that some factors will be included, although they are known to have no effect on some of the response variables.

Another example of an orthogonal array is the L_9 (3^4) design, shown in Table 6.3. The notation for this array shows that up to four factors can be used with three different levels. The subscript number nine indicates that nine trials will be needed to run the complete programme and provide analysable results.

Analysis of variance (ANOVA)

The calculations used to analyse the results of a Taguchi-designed experiment are illustrated here by way of a case study.

In recent years, car tyres have become increasingly sophisticated in their design, and the car manufacturers and motorists have

Table 6.3 Orthogonal array table, L_9

Trial	Factor			
	A	B	C	D
	Column			
	1	2	3	4
1	1	1	1	1
2	1	2	2	2
3	1	3	3	3
4	2	1	2	3
5	2	2	3	1
6	2	3	1	2
7	3	1	3	2
8	3	2	1	3
9	3	3	2	1

The interaction of columns 1 and 2 can be computed by adding the sum of squares of columns 3 and 4.

Table 6.4 Orthogonal array table, L_8

Trial	Column							Outcome
	A	B	C	D	E	F	G	
	Tread pattern	Rolling speed	Tread diameter	Surface texture	Loose surface	Wet surface	Error	Fluid retention
1	Chevron	85	Normal	Rough	27	Low	1	42
2	Chevron	85	Normal	Smooth	21	High	2	48
3	Chevron	25	Wide	Rough	27	High	2	44
4	Chevron	25	Wide	Smooth	21	Low	1	49
5	Lattice	85	Wide	Rough	21	Low	2	47
6	Lattice	85	Wide	Smooth	27	High	1	45
7	Lattice	25	Normal	Rough	21	High	1	47
8	Lattice	25	Normal	Smooth	27	Low	2	46

increasingly differentiated on the basis of tyre performance rather than just price.

One marketing-driven innovation was the idea of a tyre that had superior ability in wet conditions. The tyre manufacturer was trying to maximize the water dispersion from a tyre tread in an experimental laboratory before developing a super-dispersant premium tyre. The L_8 (2^7) array that was used is shown in Table 6.4.

Table 6.5 L_8 totals table

Factor	Level	Total	Mean
A	1	183	45.8
	2	185	46.3
B	1	182	45.5
	2	186	46.5
C	1	183	45.8
	2	185	46.3
D	1	180	45.0
	2	188	47.0
E	1	177	44.3
	2	184	47.8
F	1	177	46.0
	2	184	46.0
Error	1	183	
	2	185	
Total		368	46.0

There are six factors in this design, although there was space for seven. Each column represents a factor, and for each there are two levels. The factors are:

- A: tread pattern
- B: rolling speed
- C: tread diameter
- D: road surface texture
- E: loose surface components (dry)
- F: water/oil on surface.

For each of the eight trials there is a measure of the fluid retention. Only one repetition was performed of each trial. Later in this chapter we shall analyse more than one repetition for each trial.

Totals table

The first calculations in analysing a Taguchi orthogonal array experiment involve producing a totals table. This consists of the totals for each level of the factors in the design (Table 6.5).

Table 6.6 Sum of squares

$$S_T = (X_i)^2 + (X_{ii})^2 + \cdots + (X_n)^2 - ((\Sigma X)^2)/n)$$
$$= (42)^2 + \cdots + (46)^2 - (((368)^2)/8) = 36.00$$
$$S_A = ((\Sigma X_1 - \Sigma X_2)^2)/n$$
$$= (183 - 185)^{2/8} = 0.50$$
$$S_B = (182 - 185)^{2/8} = 2.00$$
$$S_C = (183 - 185)^{2/8} = 0.50$$
$$S_D = (180 - 188)^{2/8} = 8.00$$
$$S_E = (177 - 184)^{2/8} = 24.50$$
$$S_F = (184 - 177)^{2/8} = 0.00$$
$$S_e = (183 - 185)^{2/8} = 3.13$$

or

$$S_e = S_T - (S_A + S_B + S_C + S_D + S_E + S_F) = 0.50$$

The totals are calculated by looking at the orthogonal array, and then for each level by adding the outcomes of all trials at that level.

For example, look at the total for the first level of factor A in Table 6.5. Trials 1–4 were all carried out using this level. The outcomes of these trials were 42, 48, 44 and 49. Their sum is 183, and this is the figure recorded in the total column.

Another example is factor D at level 2. This occurred in trials 2, 4, 6 and 8. The outcomes of these trials were 48, 49, 45 and 46, respectively. The sum, recorded in the totals table, is 188.

Sum of squares

The sum of squares is calculated using the formula in the table of calculations (Table 6.6).

Degrees of freedom

The degrees of freedom are obtained for each factor, for the error term and the total. The formulae and calculations are shown in Table 6.7.

Variance

There are no changes in the way that the variance is calculated. It is always the sum of squares divided by the number of degrees of freedom.

Table 6.7 Degrees of freedom

$DF_{(Total)}$ = (No. of trials) $-$ 1 $= 8 - 1 = 7$
$DF_{(Factor)}$ = (No. of levels) $-$ 1 $= 2 - 1 = 1$
$DF_{(Error)}$ = (No. of levels) $-$ 1 $= 2 - 1 = 1$

Table 6.8 Variance

$V_A = S_A/DF_A = 0.5/1\ = 0.5$
$V_B = S_B/DF_B = 2.0/1\ = 2.0$
$V_C = S_C/DF_C = 0.5/1\ = 0.5$
$V_D = S_D/DF_D = 8.0/1\ = 8.0$
$V_E = S_E/DF_E = 24.5/1 = 24.5$
$V_F = S_F/DF_F = 0.0/1\ = 0.0$
$V_e = S_e/DF_e\ = 0.5/1\ = 0.5$

Table 6.9 F-statistics

$F_A = V_A/V_e = 0.5/0.5\ =\ 1$
$F_B = V_B/V_e = 2.0/0.5\ =\ 4$
$F_C = V_C/V_e = 0.5/0.5\ =\ 1$
$F_D = V_D/V_e = 8.0/0.5\ = 16$
$F_E = V_E/V_e = 24.5/0.5 = 49$
$F_F = V_F/V_e = 0.0/0.5\ =\ 0$

These calculations are shown in Table 6.8. V_A is the variance due to factor A and so on. V_e is the variance due to the error column.

F-statistics

Dividing the variance due to each factor by the variance of the error, gives a measure of the relative significance of each of the factors. The results of these calculations are shown in Table 6.9.

ANOVA table

As the number of factors increases, the ANOVA table becomes more useful. The initial ANOVA table for this experiment is shown in Table 6.10. It contains no new statistics, but is a convenient way of summarizing the previous calculations.

Table 6.10 Initial ANOVA table

Factor	DF	S	V	F
A	1	0.5	0.5	1
B	1	2.0	2.0	4
C	1	0.5	0.5	1
D	1	8.0	8.0	16
E	1	24.5	24.5	49
F	1	0.0	0.0	0
Error	1	0.5	0.5	
Totals	7	36		

With the increased number of factors, those that are not significant should be removed to obtain a better estimate of the relative importance of each of the significant factors. This process is known as pooling. Two pooling rules are applied in analysing the results of Taguchi-designed experiments.

First, if the F-statistic for any factor is less than 1, then pool the sum of squares for that factor into the error term. Then, if the number of factors is not approximately one-half the number of columns in the table or less, pool the factors with the smallest F-statistics until the number of factors remaining is approximately one-half the number of columns.

In this example, factors A, C and F have F-statistics less than or equal to 1. This implies that the variation due to these factors was no greater than that due to unassignable error. These factors were therefore pooled.

Although the L_8 (2^7) array could have supported seven factors, this example used only six. The pooled factors noted above represent half the original number, and so no further pooling is called for.

Pooling involves adding the sums of squares and degrees of freedom, associated with a factor, to the same statistics for the error term (Table 6.11). The pooled values are incorporated in the final ANOVA table.

The F-statistics must be recalculated using the revised error variance (Table 6.12). For the remaining factors, the pure sum of squares must be obtained (Table 6.13). The only calculation remaining before the full final ANOVA table (Table 6.14) can be prepared is that of percentage contribution. This is calculated by dividing the pure sum of squares for each factor by the total sum of squares (Table 6.15).

The interpretation of the ANOVA table begins with a check on the percentage contribution for error. This is a key measure of the success

Table 6.11 Pooling calculations

$$DF_{e(pooled)} = DF_e + DF_A + DF_C + DF_F$$
$$= 1 + 1 + 1 + 1$$
$$= 4$$

$$S_{e(pooled)} = S_e + S_A + S_C + S_F$$
$$= 0.5 + 0.5 + 0.5 + 0$$
$$= 1.5$$

$$V_{e(pooled)} = 1.5/4$$
$$= 0.375$$

Table 6.12 *F*-statistics (pooled)

$$F_B = V_B/V_e = 2/0.375 = 5.33$$
$$F_D = V_D/V_e = 8/0.375 = 21.33$$
$$F_E = V_E/V_e = 24.5/0.375 = 65.33$$

Table 6.13 Pure sum of squares

$$S_B' = S_B - (DF_B \times V_e) = 1.625$$
$$S_D' = S_D - (DF_D \times V_e) = 7.625$$
$$S_E' = S_E - (DF_E \times V_e) = 24.125$$
$$S_e' = S_e + ((DF_T - DF_e) \times V_e) = 2.625$$

Table 6.14 Final ANOVA table

Factor	DF	S	V	F	S'	P%
B	1	2.000	2.000	5.33	1.625	5
D	1	8.000	8.000	21.33	7.625	21
E	1	24.500	24.500	65.33	24.125	67
Error	4	0.375	0.375		2.625	7
Totals	7	122.875			122.875	100

$$F(95\%, 1,4) = 7.71$$
$$F(99\%, 1,4) = 21.20$$

of the experiment. This percentage contribution represents the leftover variation that was not accounted for by the factors and levels analysed in the experiment.

If the percentage contribution of the error is less than 50%, then the experiment is usually considered to have been 'good'. If the error

Table 6.15   Percentage contribution

$$P_B = S_B'/S_T = 1.625/36 \quad = 4.51\%$$
$$P_D = S_D'/S_T = 7.625/36 \quad = 21.18\%$$
$$P_E = S_E'/S_T = 24.125/36 = 67.01\%$$
$$P_e = S_e'/S_T = 2.625/36 \quad = 7.29\%$$

contributes more than 67%, then the results have not been conclusive and more work would be expected.

In this case, if we compare the F-statistics with the values tabulated in Appendix 1, then we find that factors D and E are significant. These two factors also contributed over 80% of the variation. This could be described as a particularly good experiment.

Refer to the original goal of the experiment. In this case, the object was to minimize the fluid retention. Consequently, we want to select the levels of variables that cause a statistically significant drop in the mean value for this response.

Factor D, the surface texture, has a mean fluid retention of 45.0 for level 1, and 47.0 for level 2. To minimize fluid retention level 1 is chosen.

Consider the levels of factor E, loose surface components. The first level produced a mean of 44.3 and the second level a mean of 47.8. To minimize the fluid retention, the first level of this variable would also be chosen.

Non-significant factors are used to obtain cost benefits. For factors A, B, C and F, this experiment has shown that the level chosen does not have a significant effect on the water retention. Consequently, the experimenter selects those levels that cost less to use in production. This is an important way of gaining quality improvement while reducing product cost.

We said earlier that one of the important benefits of Taguchi's approach lies in its reproducibility. Having determined optimum levels for the significant variables, we can predict the outcome that would be expected if these levels were chosen for the operating process (Table 6.16). The formula for this prediction is very straightforward.

The predicted value of 43.3 is not a guarantee that the process will deliver this value, so a confirmation run should be carried out at these levels, along with the cost-selected non-significant factors. The outcome would be expected to be close to the predicted value.

Table 6.16 Predicted optimum condition, U

$U = T'' + (D1'' - T'') + (E1'' - T'')$

where T'' is the grand mean
$D1''$ is the mean for level 1 of factor D
$E1''$ is the mean for level 1 of factor E

$U = (46) + (45 - 46) + (44.3 - 46)$
$\quad = 46 + (-1) + (-1.7)$
$\quad = 43.3$

Interactions between factors

We have analysed one of the simplest, yet most common Taguchi designs. With this example alone, you are equipped to carry out most studies for the optimization of a product or service. It really is simple; do not let yourself be put off by the unfamiliar statistics. Bear in mind that ANOVA is something that many statistics courses never reach; to have got this far is excellent.

Taguchi's designs let us look at some much more complex problems. We will now work through a more complex example, with more factors, and including some interactions.

Interactions are special relationships between factors. The interaction between two factors means that if we study one factor with the other at a constant level, we obtain one set of results, and if we then change the level of the second factor, we obtain a different set of results.

Moreover, the difference between the two sets of results is not simply additive. Interactions can often be studied by looking at a graph of the four possible combinations. Figure 6.4(a) is a graph of two factors with no significant interaction, although the two factors are significant. It is followed by a graph with a significant interaction (Figure 6.4b).

The clue to an interaction is that the two lines between the levels of the second factor are not parallel. In this case, the effect due to changing levels of factors A and B is different to the effect of changing both to the higher level.

Interactions can give extra gains to a process if they are significant. They can also be included in Taguchi designs, provided that they are identified as potentially significant beforehand. To calculate every interaction would require a large increase in the number of columns in the

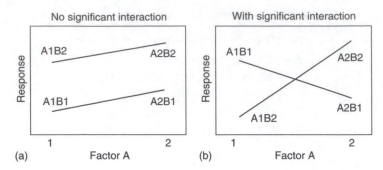

Figure 6.4 Examples of two factors without and with significant interaction

design, so only those that the experimenters believe are of likely interest are included.

In general, if you miss an interaction, the process is unlikely to go in the opposite direction. However, the optimum response may not be as large or small as possible. Either way, the effect will result in a larger error term, and this will highlight any potential problems.

Before seeing how to fit interactions into an experimental design, we will examine the L_{16} array (Table 6.17). This array can handle up to 15 factors with two levels. To produce analysable results, this requires 16 trials. The array is very similar to the L_8 array. The array is also orthogonal, and for each paired combination the number of trials is the same.

The goal of this experiment was to maximize the outflow from an effluent pipe, an important step in improving the quality of beaches in a coastal resort and enhancing the potential marketability of the area to tourists. If the outflow is not maximized, then at peak periods there can be a build-up of waste within the system; when this eventually leaves the outflow it often fails to disperse adequately, resulting in sewage reaching beaches. The chosen factors and levels are shown in Table 6.18.

Notice that there are seven factors with two levels each, and three interactions. Each interaction is treated like an additional factor, by inserting it into columns in the array. Consequently, ten columns are required, and so this experiment would not fit in an L_8 array.

The actual layout of the factors and interactions in the columns of the array is as follows. There is no order to the factors, but there is a special way in which the interactions are fitted. Table 6.19 shows how the interaction between columns of the matrix can be located for an L_{16} array. Similar tables are available for other arrays.

Table 6.17 Orthogonal array table, L_{16}

Trial	Column														
	1	2	3	4	5	6	7	8	9	10	11	12	13	14	15
1	1	1	1	1	1	1	1	1	1	1	1	1	1	2	1
2	1	1	1	1	1	1	1	2	2	2	2	2	2	1	2
3	1	1	1	2	2	2	2	1	1	1	1	2	2	1	2
4	1	1	1	2	2	2	2	2	2	2	2	1	1	2	1
5	1	2	2	1	1	2	2	1	1	2	2	1	1	2	2
6	1	2	2	1	1	2	2	2	2	1	1	2	2	1	1
7	1	2	2	2	2	1	1	1	1	2	2	2	2	1	1
8	1	2	2	2	2	1	1	2	2	1	1	1	1	2	2
9	2	1	2	1	2	1	2	1	2	1	2	1	2	1	2
10	2	1	2	1	2	1	2	2	1	2	1	2	1	2	1
11	2	1	2	2	1	2	1	1	2	1	2	2	1	2	1
12	2	1	2	2	1	2	1	2	1	2	1	1	2	1	2
13	2	2	1	1	2	2	1	1	2	2	1	1	2	2	1
14	2	2	1	1	2	2	1	2	1	1	2	2	1	1	2
15	2	2	1	2	1	1	2	1	2	2	1	2	1	1	2
16	2	2	1	2	1	1	2	2	1	1	2	1	2	2	1

Table 6.18 Maximize outflow

Factor		Level 1	Level 2
A	Stream speed	28	12
B	Channel angle	16	11
C	Pipe type	Current	New
D	Channel depth	0.05	0.50
E	Pipe material	Current	New
F	Upflow pressure	Strong	Weak
G	Effluent density	Normal	Alternative
	Interactions of interest:		
	A ∩ C	Stream speed with Pipe type	
	B ∩ D	Channel angle with Channel depth	
	D ∩ F	Channel depth with Upflow pressure	

In allocating factors to columns we start with the interactions that are to be included. Placing factor A in column 1 and factor C in column 2, the interaction table shows that the interaction between columns 1 and 2 occurs in column 3. Thus, this column is considered as occupied.

Table 6.19 Interactions between two columns, L_{16}

	2	3	4	5	6	7	8	9	10	11	12	13	14	15
1	3	2	5	4	7	6	9	8	11	10	13	12	15	14
2		1	6	7	4	5	10	11	8	9	14	15	12	13
3			7	6	5	4	11	10	9	8	15	14	13	12
4				1	2	3	12	13	14	15	8	9	10	11
5					3	2	13	12	15	14	9	8	11	10
6						1	14	15	12	13	10	11	8	9
7							15	14	13	12	11	10	9	8
8								1	2	3	4	5	6	7
9									3	2	5	4	7	6
10										1	6	7	4	5
11											7	6	5	4
12												1	2	3
13													3	2
14														1

If we place factor B in column 4, which is the next free column, then we could put factor D in column 5. However, the interaction of columns 4 and 5 is in column 1, which is occupied, and so cannot be used. Similarly, the interactions of column 4 with columns 6 and 7 occur in columns 2 and 3, respectively. Thus, they cannot be used.

Instead, placing factor D in column 8, the interaction of columns 4 and 8 occurs in column 12. With factor D in column 8, we need to find another column with a free interaction column in which to fit factor F. The interaction of columns 6 and 8 occurs in column 14. As column 14 is not in use, this is chosen.

Now that all the interactions and factors affecting them have been fitted, it remains to insert the other factors in vacant columns. Consequently, factors E and G are fitted in columns 5 and 7, respectively. Any remaining columns are ascribed to error variation.

You can see that all interaction and factor assignment has to be done before the experiment begins. From time to time the researchers discover later that there was another interaction that they would have liked to study. It is extremely unlikely (or they are very lucky) that they would be able to fit this retrospectively.

The final assignment for the effluent pipe experiment is shown in Table 6.20.

Table 6.20 Assignment of columns to factors and interactions

1	2	3	4	5	6	7	8	9	10	11	12	13	14	15
A	C	A∩C	B	E	F	G	D	err	err	err	B∩D	err	D∩F	err

Table 6.21 Alternative assignment of columns to factors and interactions

1	2	3	4	5	6	7	8	9	10	11	12	13	14	15
D	A	E	A∩C	err	C	err	B	B∩D	G	err	F	D∩F	err	err

Check yourself example: The assignment of interactions

Make sure that you have understood how factors and interactions are assigned. The example analysed in the next section comes from the same each survey, but has had its interactions assigned differently (Table 6.21). Before proceeding with the analysis, use Table 6.19 to confirm that the alternative assignment is valid.

Analysing orthogonal arrays

The analysis of an experiment of this kind proceeds in very much the same manner as that used before. First, assemble the data as shown in Table 6.22. Then calculate the total and mean for each factor and each level. Exactly the same steps should be carried out for the interaction terms, in calculating the total score of the trials with the relevant column at level 1, and similarly at level 2.

Thus, the interaction B ∩ D, in column 9, has level 1 in trial numbers 1, 3, 5, 7, 10, 12, 14 and 16. The total of the responses for these trials was 272. The same interaction, B ∩ D, has level 2 in trials 2, 4, 6, 8, 9, 11, 13 and 15. The total response in these trials was 280.

Exactly what is meant by levels of an interaction is dubious, but the calculation remains correct. The complete totals table is given in Table 6.23. Notice that we do not calculate the mean value of any interactions.

The sum of squares is also calculated for both the factors and the interactions. The calculations are given in Table 6.24.

The total degrees of freedom is the number of observations made, minus one. There were 16 observations, and so the total degrees of

Table 6.22   Maximize outflow

Trial	Column and associated factor															
	1	2	3	4	5	6	7	8	9	10	11	12	13	14	15	Outflow
	D	A	E	A∩C		C		B	B∩D	G		F	D∩F			
1	1	1	1	1	1	1	1	1	1	1	1	1	1	2	1	30
2	1	1	1	1	1	1	1	2	2	2	2	2	2	1	2	36
3	1	1	1	2	2	2	2	1	1	1	1	2	2	1	2	44
4	1	1	1	2	2	2	2	2	2	2	2	1	1	2	1	42
5	1	2	2	1	1	2	2	1	1	2	2	1	1	2	2	26
6	1	2	2	1	1	2	2	2	2	1	1	2	2	1	1	28
7	1	2	2	2	2	1	1	1	1	2	2	2	2	1	1	40
8	1	2	2	2	2	1	1	2	2	1	1	1	1	2	2	34
9	2	1	2	1	2	1	2	1	2	1	2	1	2	1	2	36
10	2	1	2	1	2	1	2	2	1	2	1	2	1	2	1	32
11	2	1	2	2	1	2	1	1	2	1	2	2	1	2	1	36
12	2	1	2	2	1	2	1	2	1	2	1	1	1	2	2	46
13	2	2	1	1	2	2	1	1	2	2	1	1	2	2	1	40
14	2	2	1	1	2	2	1	2	1	1	2	2	1	1	2	28
15	2	2	1	2	1	1	2	1	2	2	1	2	1	1	2	28
16	2	2	1	2	1	1	2	2	1	1	2	1	2	2	1	26

freedom is 15. For each factor, there is one less degree of freedom than the number of levels used. In this case all factors were at two levels, and so each has one degree of freedom.

For each interaction, the number of degrees of freedom is the product of the degrees of freedom for each of the constituent factors. For an experiment where all factors are at two levels, this will be: $(2 - 1) * (2 - 1) = 1$.

The calculations for degrees of freedom are given in Table 6.25.

The initial ANOVA table is created by calculating the variance for each factor and interaction, and then calculating the F-statistic. The formulae for these are:

$$V_A = S_A/DF_A$$

$$F_A = V_A/V_e$$

The initial ANOVA table is shown in Table 6.26. Studying the table shows that factors B, F, D and E and the interaction B ∩ D have F-statistics of less than one, and are thus marked for pooling. For an L_{16} array, leaving less than half the columns would mean seven or eight factors. As there are five left over, this meets the requirement.

Table 6.23 L_{16} totals table

Factor	Level	Total	Mean
$D \cap F$	1	256	32.00
	2	296	37.00
G	1	262	32.75
	2	290	36.25
$A \cap C$	1	256	32.00
	2	296	37.00
A	1	302	37.75
	2	250	31.25
F	1	280	35.00
	2	272	34.00
D	1	280	35.00
	2	272	34.00
C	1	262	32.75
	2	290	36.25
$B \cap D$	1	272	34.00
	2	280	35.00
B	1	280	35.00
	2	272	34.00
E	1	274	34.25
	2	278	34.75
Totals		552	34.50

Table 6.24 Sum of squares

$$S_{total} = (30^2) + \cdots + (26^2) - \{(552^2)/16\} = 644$$
$$S_{D \cap F} = ((256 - 296)^2)/16 = 100$$
$$S_G = ((262 - 290)^2)/16 = 49$$
$$S_{A \cap C} = ((256 - 296)^2)/16 = 100$$
$$S_A = ((302 - 250)^2)/16 = 169$$
$$S_F = ((280 - 272)^2)/16 = 4$$
$$S_D = ((280 - 272)^2)/16 = 4$$
$$S_C = ((262 - 290)^2)/16 = 49$$
$$S_{B \cap D} = ((272 - 280)^2)/16 = 4$$
$$S_B = ((280 - 272)^2)/16 = 4$$
$$S_E = ((274 - 278)^2)/16 = 1$$
$$S_{err} = SS - (\text{Sum of individual factors SS}) = 160$$

Table 6.25 Degrees of freedom

Total degrees of freedom	$= 16 - 1 = 15$
Factor degrees of freedom	$= 2 - 1 = 1$
Interaction DF	$=$ Factor 1 DF $\times$ Factor 2 DF
	$= (2 - 1) \times (2 - 1)$
	$= 1$
Error DF	$=$ Total DF $-$ (Sum of all other factors' DFs)
	$= 15 - (10)$
	$= 5$

Table 6.26 Initial ANOVA table

Source	DF	S	V	F
D	1	4	4	0.125
A	1	169	169	5.281
E	1	1	1	0.031
A ∩ C	1	100	100	3.125
C	1	49	49	1.531
B	1	4	4	0.125
B ∩ D	1	4	4	0.125
G	1	49	49	1.531
F	1	4	4	0.125
D ∩ F	1	100	100	3.125
Error	5	160	32	
Totals	15	644		

The pooling process simply consists of adding the identified factors sums of squares and degrees of freedom to the error values. The F-statistics, pure sum of squares and percentage contributions are calculated, and the final ANOVA table is prepared (Table 6.27).

Note that the percentage contribution for the error term is 41.2%, indicating that the experiment has been reasonably successful in identifying the major sources of variation.

One factor, A, and two interactions, D ∩ F and A ∩ C, have percentage contributions that represent between them almost 50% of the variation.

The F-statistics show that all the remaining factors except G were statistically significant at a 95% confidence level, so we can expect to achieve a substantial improvement by incorporating them into the optimization process.

Table 6.27 Final ANOVA table

Source	DF	S	V	F	S′	P%
D ∩ F	1	100	100	5.6	82.3	12.8
G	1	49	49	2.8	31.3	4.9
A ∩ C	1	100	100	5.6	82.3	12.8
A	1	169	169	9.5	151.3	23.5
C	1	49	49	2.8	31.3	4.9
Error (pooled)	10	177	17.7		265.5	41.2
Totals	15	644			644	100

Table 6.28 Mean interaction levels

Mean of ...
D1F1 = 33
D2F1 = 37
D1F2 = 37
D2F2 = 31

A1C1 = 33.5
A2C1 = 32
A1C2 = 42
A2C2 = 30.5

To maximize the effluent outfall, we refer to the original totals table, and select those levels that have the largest mean values. These would be A1 and C2. However, there were also two significant interaction terms, D ∩ F and A ∩ C.

To obtain the mean levels for the interaction terms, select the four trials that have the same combinations of levels and calculate the mean for these trials. The four such values are shown in Table 6.28 and represented graphically in Figure 6.5.

To obtain the predicted response at the optimum levels, we previously used the grand mean, and added the additional contributions from the mean levels of significant factors. In this case, as A and C were also present in the interaction they are not included, otherwise we would be 'double accounting'.

In this example, we use the same procedure for the factor that was not in the interaction term. For the interaction term, and those factors that were included in it, yet were also significant in their own right, we use

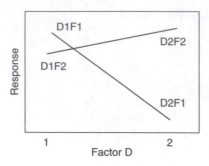

Figure 6.5 Mean levels for the interaction terms

Table 6.29 Predicted optimum response

$$U = T'' + (A1C2'' - T'') + (D2F1'' - T'') \text{ or } (D1F2'' - T'')$$
$$\text{(depending on cost)}$$
$$U = 34.5 + 7.5 + 2.5$$
$$U = 44.5$$

the mean level of the interaction term that contributed the largest mean level (Table 6.29).

The handling of interactions represents one of the tradeoffs in using this form of experimental design. With full factorials it is possible to include all the interaction terms, but often at an enormous cost, because to do so involves substantially more experimental trials. With Taguchi designs, experimenters can use their own expertise to decide which interactions are worth including, and which they would find difficult to interpret. In this way they can achieve substantial reductions in cost and in time taken to complete the experimental programme.

In general, interactions are rarely as significant as the overall main factors in an experiment.

Analysing orthogonal arrays with repetitions

The analyses considered so far have all involved one repetition. That is, the response factor has been measured only once for each trial. In most experiments, however, several measurements are made of the same

Table 6.30 Minimize viscosity

Design	Factor	Level 1	Level 2
A	Mixing temperature	50°	54°
B	Packing temperature	8.5°	13.0°
C	Packing type	150 seconds	200 seconds
D	Stabilizer type	Type A	Type B
	Interactions:		
	D ∩ C		
	B ∩ C		

Experiment and results

Trial	1 A	2 B	3 D ∩ C	4 C	5 err	6 B ∩ C	7 D	Viscosity values (repetitions)					Total
1	1	1	1	1	1	1	1	8	7	6	5	5	31
2	1	1	1	2	2	2	2	3	4	4	4	5	20
3	1	2	2	1	1	2	2	8	7	8	12	10	45
4	1	2	2	2	2	1	1	8	9	7	9	8	41
5	2	1	2	1	2	1	2	9	9	8	10	11	47
6	2	1	2	2	1	2	1	8	7	7	10	8	40
7	2	2	1	1	2	2	1	8	9	8	13	13	51
8	2	2	1	2	1	1	2	8	6	8	6	8	36

response for each trial. This dramatically improves the accuracy of the work.

This case study shows how to handle the data generated when more than one repetition is carried out.

The marketing department of a large household products company had been conducting research with consumers. The feedback they received was that some of their foam cleaners were too sticky, making it hard for people to wipe over a work-surface with them and causing the dispensers to clog up.

The problem they presented to the production team therefore concerns minimizing the viscosity of one such domestic foam cleaner. There are four factors at two levels, and there are two interactions that the engineers have suggested that may be important (Table 6.30).

The first stage in the analysis is to create a totals table. This is done in the same manner as before, but here there are five repeats of each trial. Thus, the total for level 1 of factor C is the total of all the measurements

Table 6.32   Sum of squares

S_T	$= (8^2) + (7^2) + \cdots + (8^2) - \{(311^2)/40\} = 207$	
S_A	$= ((137 - 174)^2)/40$	$= 34$
S_B	$= ((138 - 173)^2)/40$	$= 31$
S_C	$= ((174 - 137)^2)/40$	$= 34$
S_D	$= ((163 - 148)^2)/40$	$= 6$
$S_{D \cap C}$	$= ((138 - 173)^2)/40$	$= 31$
$S_{B \cap C}$	$= ((155 - 156)^2)/40$	$= 0$
S_{e1}	$= ((152 - 159)^2)/40$	$= 1$
S_{e2}	$= $ Total SS $-$ (Sum of all others)	$= 70$

Table 6.33   Initial ANOVA table

Source	DF	S	V	F
A	1	34	34	11
B	1	31	31	10
C	1	34	34	11
D	1	6	6	2
D ∩ C	1	31	31	10
B ∩ C	1	0	0	0
e1	1	1	1	
e2	32	70	2	
Total	39			

The reasoning behind this is that the between-experiment error is the term that we would like to test against. The within-experiment error would ideally be small (no variation between trials would be wonderful).

The pooling rules determine whether the within-experiment variation is so great that we have no choice but to include it. In this case, the primary error is less than the secondary, so we have no alternative but to pool the terms for error sum of squares and degrees of freedom by simple addition.

From the initial ANOVA table, we decide which factors should be further pooled in the error terms. First, we look for terms that have an F-statistic of less than one. This means that the interaction B ∩ C will be pooled.

Of the eight columns there are still five factors or interactions, and so the one with the next smallest F-statistic is pooled. This is factor D.

Table 6.34   Final ANOVA table

Source	DF	S	V	F	S	P%
A	1	34	34	17	32	15
B	1	31	31	15	29	14
C	1	34	34	17	32	15
D ∩ C	1	31	31	15	29	14
e(pool)	35	77	2		85	41
Total	39	207				

$F(95\%, 1,35) = 4.12$
$F(99\%, 1,35) = 7.41$

Table 6.35   Mean interaction levels

$D1C1'' = 8.2$
$D1C2'' = 8.1$
$D2C1'' = 9.2$
$D2C2'' = 5.6$

Table 6.36   Prediction of optimum response

$U = T'' + (A1'' - T'') + (B1'' - T'') + (D2C2'' - T'')$
$= 7.8 + (-0.9) + (-0.9) + (-2.2)$
$= 3.8$

Recalculating the variances and F-statistics, and calculating the pure sums of squares and percentage contributions, we prepare the final ANOVA table (Table 6.34).

The object of the experiment was to minimize the viscosity. Consequently, the levels of the significant factors that have the lowest mean values in the totals table are the ones that should be chosen.

For the interaction term, the mean levels of each of the four possible combinations are calculated, and then the lowest selected (Table 6.35). The lowest of these values is that for D2C2', so this is chosen.

The overall optimum condition is with the first levels of A and B, and with the D2C2 combination. As C is included in the interaction term, we do not use the factor C alone in calculating the expected optimum response. The overall success of the programme will be judged on the closeness of the prediction to the confirmatory trial result (Table 6.36).

Table 6.37 Example with no error term

Column	Factor							Outcome
	A	B	C	D	E	F	G	
1	1	1	1	1	1	1	1	176
2	1	1	1	2	2	2	2	202
3	1	2	2	1	1	2	2	187
4	1	2	2	2	2	1	1	204
5	2	1	2	1	2	1	2	197
6	2	1	2	2	1	2	1	191
7	2	2	1	1	2	2	1	197
8	2	2	1	2	1	1	2	193

Analysing orthogonal arrays without error terms

Orthogonal arrays can only have no error terms when there is only one run of each experimental trial (i.e. there are no repeats), and all the columns of the array are occupied by factors or interactions. Table 6.37 is an L_8 (2^7) array that has seven factors (A–G), with only one repetition for each experimental trial (i.e. no repeats).

The analysis of this experiment begins with the calculation of the sums of squares for each of the factors. The formula used for this calculation is exactly the same as before:

$$S_A = (\Sigma Y_1 - \Sigma Y_2)^2/N$$

There is no sum of squares for the primary error because there are no vacant columns in the array. Similarly, there is no sum of squares for the secondary error, as none of the trials has been repeated. The sums of squares calculations are shown in Table 6.38.

Once the sums of squares have been calculated, the initial ANOVA table can be drawn up (Table 6.39). This will have sums of squares, degrees of freedom and variance, but no F-statistics, as there is no variance term for error at this stage.

To form the final ANOVA table, the two rules of pooling described earlier are used. As there are no F-statistics we cannot use F-ratios of less than one to pool terms, so we have to rely on the second rule.

There are seven factors in the original experiment, and so this needs to be reduced to about three or four. By pooling factors A, C, F and G, the number of factors is reduced to three. These pooled factors are chosen because they have the lowest variances in the initial ANOVA table.

The sums of squares and degrees of freedom terms are pooled for these factors, and the final ANOVA table is drawn up (Table 6.40).

The analysis would then be completed in the usual way, with the optimum levels being determined and a predicted optimum condition being

Table 6.38   Sum of squares

$$S_A = (769 - 778)^2/8 \quad = \quad 10$$
$$S_B = (766 - 781)^2/8 \quad = \quad 28$$
$$S_C = (768 - 779)^2/8 \quad = \quad 15$$
$$S_D = (757 - 790)^2/8 \quad = 136$$
$$S_E = (747 - 800)^2/8 \quad = 351$$
$$S_F = (770 - 777)^2/8 \quad = \quad 6$$
$$S_G = (768 - 779)^2/8 \quad = \quad 15$$
$$S_{e1} = \text{No vacant columns} = \quad 0$$
$$S_{e2} = \text{No repetitions} \quad = \quad 0$$

Table 6.39   Initial ANOVA table

Source	DF	S	V	Pool
A	1	10	10	x
B	1	28	28	
C	1	15	15	x
D	1	136	136	
E	1	351	351	
F	1	6	6	x
G	1	15	15	x
Total	7	561		

Table 6.40   Final ANOVA table

Source	DF	S	V	F	S'	P%
B	1	28	28.0	2.4	16.5	3
D	1	136	136.0	11.8	124.5	22
E	1	351	351.0	30.5	339.5	61
e(pool)	4	46	11.5		80.5	14
Total	7	561			561	100

estimated. This prediction would then be tested through a confirmation trial, and the success of the experiment would be determined.

Orthogonal arrays with no primary error term

Another problem that can occur is when there are repetitions but no spare columns. Initially, there will be no primary error term. We have already seen one solution to this difficulty, where the secondary error term forms the F-statistics.

Alternatively, the factors with the smallest variances can be pooled together to form a primary error term. A decision can then be made on which is the best error term for further analysis.

The example below of an L_4 array is used to illustrate this point. All three columns are occupied by factors (A–C), but there are two repetitions for each experimental trial. Consequently, there will be no primary error term, but there will be a secondary error (Table 6.41).

To begin with, the sums of squares are calculated, using the familiar formula (Table 6.42), $S_A = (\Sigma Y_1 - \Sigma Y_2)^2/N$.

The next stage is to determine the number of degrees of freedom associated with each factor and error term. The total degrees of freedom

Table 6.41 L_4 array with no error but with repetitions

Column	Factor			Thickness	
	A	B	C		
1	1	1	1	22	27
2	1	2	2	30	36
3	2	1	2	29	25
4	2	2	1	30	29

Table 6.42 Sum of squares

$S_A = (115 - 113)^2/8$	$= 0.5$
$S_B = (103 - 125)^2/8$	$= 60.5$
$S_C = (108 - 120)^2/8$	$= 18$
$S_T = 6616 - (228^2)/8$	$= 118$
$S_{e1} = $ No vacant columns	$= 0$
$S_{e2} = $ Total SS $-$ (Sum of individuals) $= 39$	

Table 6.43 Initial ANOVA table

Source	DF	S	V	Pool
A	1	0.5	0.5	x
B	1	60.5	60.5	
C	1	18.0	18.0	
e1	0	0.0	0.0	
e2	4	39.0	9.8	
Total	7	118		

Table 6.44 Initial ANOVA table with primary error formed

Source	DF	S	V
B	1	60.5	60.5
C	1	18	18
e1(pri)	1	0.5	0.5
e2(sec)	4	39	9.75
Total	7	118	

can be calculated as one less than the total number of observations (i.e. $8 - 1 = 7$). The factors were all studied at two levels, and so each has one degree of freedom associated with it (i.e. $2 - 1 = 1$).

There are no degrees of freedom associated with the primary error, as there are no vacant columns. The number of degrees of freedom due to the secondary error term is calculated by deduction (i.e. 7 minus 1 for each factor, $7 - 3 = 4$).

The initial ANOVA table can now be formed by pooling the factors with the smallest variances, again using the second pooling rule (Table 6.43).

For this experiment, since factors B and C have variances so much greater than that of the other factor, A, it will be pooled.

The sum of squares and degrees of freedom for A form a primary error term. The revised initial ANOVA table is shown in Table 6.44.

As the primary error variance is less than that of the secondary error, the two are added to form the F-statistics. If it were greater, the secondary terms would be discarded in the final ANOVA table shown in Table 6.45.

The percentage contribution for factor B is only 43%, and that of the error term is 57%. This indicates that the experiment is not very good

Table 6.45 Final ANOVA table before and after pooling non-significant factors

Source	DF	S	V	F
B	1	60.5	60.5	7.66
C	1	18.0	18.0	2.29
e	5	39.5	7.9	
Total	7	118.0		

Factor B is significant at 95% confidence.
Factor C is not significant.

Source	DF	S	V	F	S'	P%
B	1	60.5	60.5	6.3	50.9	43
e	6	57.5	9.6		67.1	57
Total	7	118.0			118.0	

when analysing the central tendency of the process. The optimum level of factor B would be selected, a predicted optimum condition would be obtained and a confirmation trial would be carried out to verify the results. Further experiments would be necessary.

Analysing orthogonal arrays with unequal sample sizes

Experiments with continuous variable data

Rather too often we find that the experiment suffers from problems during some of the trials. Consequently, we may end up with some trials with several repeats, and others with only one result.

The formula that was originally used for two-level factors does not work when there is an unequal sample size. Instead, we must use a formula very similar to the one used with three-level experiments. This formula is shown in Table 6.46.

Experiments with attribute variable data

When using attribute data and confronted with the same problem of incomplete repetitions, a similar change is made to the calculation for

Table 6.46   Sum of squares for unequal sample sizes

Two-level factor: variable data

$$\text{Sum of squares of factor A} = \frac{(\Sigma A_1)^2}{n_1} + \frac{(\Sigma A_2)^2}{n_2} = \frac{(\Sigma A_1 + \Sigma A_2)^2}{N}$$

where ΣA_1 = Total of outcomes for first level of factor A
ΣA_2 = Total of outcomes for second level of factor A
n_1 = No. of observations for first level of factor A
n_2 = No. of observations for second level of factor A
N = Total no. of observations

Table 6.47   Sum of squares for unequal sample sizes

Two-level factor: attribute data

Sum of squares of factor A = Contribution due to class I + Contribution due to class II

$$\left[\frac{(\Sigma A_1)^2}{n_1} + \frac{(\Sigma A_2)^2}{n_2} - \frac{(\Sigma A_1 + \Sigma A_2)^2}{N} \right] \times W_1$$

$$+$$

$$\left[\frac{(\Sigma A_1)^2}{n_1} + \frac{(\Sigma A_2)^2}{n_2} - \frac{(\Sigma A_1 + \Sigma A_2)^2}{N} \right] \times W_2$$

where ΣA_1 = Total of outcomes for first level of factor A for that class
ΣA_2 = Total of outcomes for second level of factor A for that class
W_1 = Weight for class I
W_2 = Weight for class II
n_1 = No. of observations for first level of factor A
n_2 = No. of observations for second level of factor A
N = Total no. of observations

sums of squares. The alternative version of the formulae for sums of squares is given in Table 6.47.

Non-linear factors and three-level designs

So far, we have only discussed experiments using two-level factors. Some changes in the calculations are required when an increased number of levels is used. A three-level experiment using the L_9 orthogonal array will be used to illustrate these differences.

This experiment was carried out in the engine laboratory of a petrochemical company. Working with a demounted diesel engine, they wanted

Table 6.48 Experimental design and results

	Factor				Level 1	Level 2	Level 3
A	Torque				−5	0	5
B	Air filter				0	1	2
C	Cooler setting temperature				−5	0	5
D	Muffler type				Normal	New #1	New #2

Trial	Factor				Consumption values			Totals
	A	B	C	D				
1	1	1	1	1	4	4	4	12
2	1	2	2	2	3	3	3	9
3	1	3	3	3	2	2	3	7
4	2	1	2	3	3	3	2	8
5	2	2	3	1	3	3	3	9
6	2	3	1	2	3	3	3	9
7	3	1	3	2	2	2	2	6
8	3	2	1	3	2	2	3	7
9	3	3	2	1	4	4	4	12
							Total	79

to optimize fuel consumption. There was considerable resistance to the idea that factors were linear, so a pilot experiment was run using three-level factors (Table 6.48).

The main change to the analysis when three levels are studied is in the calculation of the sums of squares (Table 6.49).

For this design there is no primary error, as there were no free columns in the array. The secondary error term is obtained by deduction. This formula can be applied to any number of levels. There are no differences in the general formula for the calculation of degrees of freedom.

The initial ANOVA table can be prepared in exactly the same way (Table 6.50).

Examining this ANOVA table, there are two factors with F-statistics of less than one, namely A and B. The pooled error term is obtained by adding the sum of squares and the degrees of freedom for the two pooled factors to the already established error term. The final ANOVA table is illustrated in Table 6.51.

The percentage contribution for error was 26%, indicating a good experiment. Factors D and C were statistically significant and had

Table 6.49 Sum of squares

General three-level formula
$$= (*1^2)/N1 + (*2^2)/N_2 + (*3^2)/N_3 - ((*1 + *2 + *3)^2)/N$$

Example

$$S_A = (28^2)/9 + (26^2)/9 + (25^2)/9 - (79^2)/27 = 0$$
$$S_B = (26^2)/9 + (25^2)/9 + (28^2)/9 - (79^2)/27 = 0$$
$$S_C = (28^2)/9 + (29^2)/9 + (22^2)/9 - (79^2)/27 = 3$$
$$S_D = (33^2)/9 + (24^2)/9 + (22^2)/9 - (79^2)/27 = 8$$
$$S_T = (41^2) + (38^2) + \cdots + (40^2) - (79^2)/27 = 14$$
$$S_{e2} = 14 - (0 + 0 + 3 + 8) = 3$$

Table 6.50 Initial ANOVA table

Source	DF	S	V	F
A	2	0	0	0
B	2	0	0	0
C	2	3	1.5	7.5
D	2	8	4	20
e2	18	3	0.2	
Total	26	14		

Table 6.51 Final ANOVA table

Source	DF	S	V	F	S'	P%
C	2	3	1.5	10.7	2.7	19
D	2	8	4	28.6	7.7	55
Error	22	3	0.14		3.6	26
Total	26	14			14	100

percentage contributions of 55% and 19%, respectively. The mean levels of the two significant factors are shown in Table 6.52.

The object of the experiment was to minimize consumption values. The levels with the lowest mean values should be used for the prediction of the optimum response. These are the third levels of each of the two factors.

An advanced method that can be used with three levels is to compute a linear and quadratic component for the significant factors. This would

Table 6.52   Mean levels for significant factors

$C1'' = 3.1$
$C2'' = 3.2$
$C3'' = 2.4$

$D1'' = 3.7$
$D2'' = 2.7$
$D3'' = 2.4$

help to determine whether the significant effect was due to a linear relationship of levels or a quadratic relationship of levels. For factors that are not measurable variables but categorical, such a process is meaningless.

Multiple characteristics and mixed level designs

So far, we have concentrated on only one response at a time. Usually, though, we will measure more than one. The objective of an experiment will generally be to optimize one response factor, but at the same time to understand the behaviour of others. Changes in a level may provide a beneficial change in one response, but a deleterious one in others. Consequently, all responses for a particular system should be studied, not just the one in need of optimization.

To evaluate multiple responses, an ANOVA is performed for each response independently. From the results of the final ANOVAs, a decision matrix is formed. This lists the different 'quality' characteristics, or response factors, horizontally. Then, in vertical columns, the different factors in the experiment are listed. The matrix consists of boxes for each control and response factor. In each box, we record whether the factor was significant or not, and the best level for that factor, along with its percentage contribution.

When deciding which level of a control factor should be chosen, we take into account the improvement that will come from using a particular level, and the cost of achieving that level. Thus, there are four possible combinations of cost and benefit (Table 6.53).

If a factor does not significantly affect the response, then we would normally use the level of that factor that costs least (option 4). When a factor significantly affects the response, and the optimum level is also the cheapest, we would expect to use that level (option 2).

147

Table 6.53 Evaluating the results

1 Factor significant: Cost higher
2 Factor significant: Cost lower
3 Factor insignificant: Cost higher
4 Factor insignificant: Cost lower

Table 6.54 Optimum responses

Factor	Responses			
	Cold solders	Bridges	Cracks	Current loss
Supplier				
Smalls	✓			✓
Patak	Smalls (13%)			Patak (29%)
Preheat temperature				
270	✓	✓	✓	
410	270 (24%)	270 (23%)	410 (12%)	
Conveyor speed				
4	✓	✓		
6	4 (38%)	4 (17%)		
Solder pot temperature				
410		✓		✓
580		410 (33%)		580 (27%)
Vibration				
Low damping				
High damping				

This example comes from an unusual electronics assembly process in India. The purpose of the experiment was to reduce the current loss for part of a circuit. This outcome was chosen because it represented the greatest cost of quality. The control factors were: supplier, preheat temperature, conveyor speed, solder pot temperature, flux density and vibration setting. The responses that were assessed involved: number of cold solders, number of bridges, number of visible cracks and current loss. The optimum responses are shown in Table 6.54. For each factor we look at the effect on each quality characteristic.

- *Supplier*: using Patak as a vendor will improve the current loss of the circuit, with a percentage contribution of 29%. Using Smalls will improve the number of cold solders, with a percentage contribution of 11%. Since the highest cost of quality was for current loss, Patak should be used to obtain an overall improvement in the process when evaluating both quality and cost.

■ *Preheat*: using level 1, which was 270°F, gave an improvement in the number of no solders and in the number of bridges. Choosing 410°F improved (i.e. reduced) the number of visible cracks. However, the percentage contribution for visible cracks was smaller than for the numbers of bridges and cold solders. Consequently, it would be concluded that the first level of preheat temperature, 270°F, should be used.

■ *Conveyor speed*: there is no significant effect on visible cracks or current loss. There is a significant effect on cold solders and on bridges, both being optimal at the lower level of the factor. As a result, unless the cost is too great, the slower conveyor speed of four feet per minute should be used.

■ *Solder pot temperature*: the first level of 410°F gave an improvement in the number of bridges, with a percentage contribution of 33%. The experiment also showed that using the second level of 580°F improved current loss values, with a percentage contribution of 27%. Although the percentage contribution of 27% is less than that for the improvement in number of bridges, the 580°F level should be used, as the original objective of the experiment was to optimize the current loss of the circuit.

■ *Vibration setting*: as the vibration setting had no effect on the quality characteristics, the most cost-effective level should be chosen. In this case, running at low damping improved the cost benefits of the production process, and so this level was selected.

Once a suite of levels has been chosen, the optimum performance of the quality characteristics should be individually predicted. If the actual improvements expected from the set of levels selected are not substantial, or not cost-effective, it might be decided to re-evaluate the matrix.

Evolutionary operations

The concept of evolutionary operations (EVOP) was developed by George Box, and has been available since 1957 (Box and Draper, 1969).

Table 6.55 Evolutionary operations: number of cycles to detect significant changes

Ratio of detection size to standard deviation	0.84	1.13	1.41	1.70	2.00	2.25
No. of runs (excluding centre)	62	36	24	18	14	12
No. of cycles (two factors)	15	9	6	5	4	3
No. of cycles (three factors)	8	4.5	3	2	2	2

Unlike most other experimental design techniques, including Taguchi's methods, it is less concerned with research and development than with improvement. Like Taguchi's methods, it is eminently practical.

Many plants have been operating under similar conditions for a very long time. This set of conditions probably arose for a mixture of technical, behavioural and economic reasons. As these conditions change, processes become progressively less efficient. EVOP is a simple technique to put the system back in line.

- EVOP nudges a plant towards optimum conditions without causing dramatic disturbances or catastrophic cutbacks in production.
- EVOP is the only legitimate form of chasing the variable.
- EVOP programmes have three basic steps: (1) systematic small changes are introduced in the levels of operating variables; (2) information is collected and summarized, often posted so that plant operatives can see what is happening; (3) further changes are made, and the process is repeated.

Minimum cycle numbers

Table 6.55 shows how many cycles are required to detect effects of size, D, with a detection rate of 90%, using a significance level of 95%, when the background standard deviation is S.

Evolutionary operation for a two-factor design

- Select an appropriate experimental region around the existing operating conditions. Decide on the acceptable deviations in the two parameters.
- Perform the five trials in the order shown in Table 6.56. Do not randomize the sequence.

Table 6.56 Evolutionary operation for two factors

Condition	Factor A	Factor B
0	0	0
1	−	−
2	+	+
3	+	−
4	−	+

Factor A

```
Factor A │
         │    4              2
         │          0
         │    1              3
         └──────────────────────
                              Factor B
```

- Repeat the cycle for the necessary number of times to detect a significant effect.
- Average the results for each trial condition.
- Follow the direction of optimality, defining the next region for investigation and choosing levels of the two factors with similar sizes of change.
- The results of a complete cycle can be analysed using a simple two-factor ANOVA.
- Repeat the pattern for the next phase.

Evolutionary operation for a three-factor design

- Select an appropriate experimental region around the existing operating conditions. Decide on acceptable deviations in the three parameters.
- Perform the nine trials in the order shown in Table 6.57. Do not randomize the sequence.
- Repeat the cycle for the necessary number of times to detect a significant effect.
- Average the results for each trial condition.
- Follow the direction of optimality, defining the next region for investigation, and choosing levels of the three factors with similar sizes of change.
- The results of a complete cycle can be analysed using a simple three-factor ANOVA.
- Repeat the pattern for the next phase.

Table 6.57   Evolutionary operation for three factors

Trial	Factor A	Factor B	Factor C
0	0	0	0
1	−	−	−
2	+	+	−
3	+	−	+
4	−	+	+
5	+	−	−
6	−	+	−
7	−	−	+
8	+	+	+

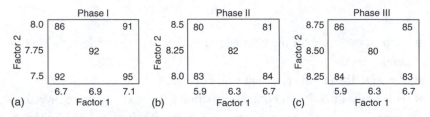

Figure 6.6   Results of an evolutionary operations experiment

Evolutionary operations example

This example is concerned with minimizing the cost per tonne of product where two variables are important:

- reflux ratio for a distillation column.
- recycle flow to purge flow ratio.

The *D/S* ratio required five cycles per phase. The five points are each run five times, and the results shown in Figure 6.6 are the average of the five.

Accumulation analysis

So far, we have considered only data of the variable or continuous type, i.e. measurement data. Often the data gathered in an experiment are not measured, but are categorized and referred to as attribute data. We look at attribute data in further detail in Chapter 7. To analyse this in the Taguchi methodology, a tool called accumulation analysis is used.

To use accumulation analysis, the data must be categorized into three or more classes. These must also have some order to them; for example, three grades of product: good, normal and bad.

The orthogonal array is still used to produce the necessary trial sequences. Interactions are handled in the same way. The sum of squares, however, is computed differently, as are the degrees of freedom. Taguchi's signal-to-noise concepts cannot be used with accumulation analysis.

Accumulation analysis with only one repetition

Anyone who has used a drill to make a hole in wood will appreciate this problem. As the drill cuts it creates heat and under certain circumstances this is sufficient to create a burnt surface on the wood.

A furniture manufacturer using semi-automated jigs wanted to establish the safest operating conditions for his machines. The experiment generated only one data value for each trial by carrying out only one repetition, i.e. all trials are performed only once. The response being assessed was tendency to burn, and this was classified in three categories: very little, some smoke and charring. For each experimental trial, a set of conditions was fixed and the product was examined. It was then decided which of the three categories was the most appropriate description of the product.

For this problem, an L_8 array was used. Factors and interactions were assigned as usual. These were fitted into the array using the interaction table that accompanies most of the Taguchi designs. The original data taken for each trial are shown in Table 6.58.

Data accumulation

The process known as accumulation is used to analyse data of this kind. Three classes are formed:

- class I: the frequency of the very little category
- class II: the frequency of the very little and some smoke categories

Table 6.58 Factor alignment in the L_8 array with data

Trial	Factor						Error	Grade		
	A	B	C	D	E	F	G	VL	SS	C
1	1	1	1	1	1	1	1	1	0	0
2	1	1	1	2	2	2	2	1	0	0
3	1	2	2	1	1	2	2	0	0	1
4	1	2	2	2	2	1	1	0	0	1
5	2	1	2	1	2	1	2	0	0	1
6	2	1	2	2	1	2	1	0	0	1
7	2	2	1	1	2	2	1	1	0	0
8	2	2	1	2	1	1	2	0	1	0
							Totals	3	1	4

- class III: the frequency of the very little, some smoke and charring categories.

For example, in trial 8, the result was described as some smoke. Class I has a value of 0, as none of the results fell into the very little bin. Class II has a value of 1, because the holes fell into the some smoke category, while class III also has a value of 1, as it is the accumulation of very little, some smoke and charring results.

Provided that the data have been tabulated with the lowest grade to the left and the highest to the right, then accumulation will be from left to right.

It is not really as complicated as it may sound: try following through a few rows of the L_8 array to check that you are calculating the accumulation figures correctly. The accumulated frequencies are shown in Table 6.59.

Calculation of proportions

First, we calculate a proportion for each of the classes, except the last. The last class is not calculated, as it will always have the same total as the number of trials, and its sum of squares will always be zero.

For each remaining class, divide the class frequency by the total number of observations in the experiment. Thus, for class I in this example there were three data points. The total number of observations is eight,

Table 6.59 L$_8$ data accumulated

	VL	SS	C	I	II	III
1	1	0	0	1	1	1
2	1	0	0	1	1	1
3	0	0	1	0	0	1
4	0	0	1	0	0	1
5	0	0	1	0	0	1
6	0	0	1	0	0	1
7	1	0	0	1	1	1
8	0	1	0	0	1	1
Total	3	1	4	3	4	8

Table 6.60 Proportions

For every class except the last:

$$P_i = \frac{\text{Frequency in class}}{\text{Total no. of observations}}$$

so the proportion in class I is 3/8 or 0.38. The proportion falling in class II is four observations out of eight, or 0.5. Table 6.60 shows the formula for calculating proportions.

Calculation of weightings

The next step is to calculate weightings for each class except the last. In the simplest case, a weighting is calculated for the first two classes. This is a standard statistical procedure, used to allow for the extremes that are found away from the norm. The midpoint of the frequencies is four (out of eight), and the weighting will be greater the farther a class is away from the midpoint.

The weighting is calculated by taking the proportion for that class multiplied by one minus the proportion, and then calculating the reciprocal of the result (Table 6.61).

In this case, the weight of class I will be the reciprocal of 3/8 times 1 minus 3/8. This can be calculated out to 4.27. Similarly, the weight of class II is the reciprocal of 4/8 times 1 minus 4/8, which works out at 4.00.

Table 6.61 Weightings

For every class except the last:

$$W_i = 1/(P_i \times (1 - P_i))$$

Table 6.62 Total sum of squares

$$S_T = \text{(Total no. of observations)} \times \text{(No. of classes} - 1)$$

Table 6.63 Sum of squares for individual factors

$$S_i = (S_{i1} \times W_1) + (S_{i2} \times W_2)$$

where

$S_{i1} = ((T_{i1} - T_{i2})^2)/\text{(Total no. of observations)}$
$T_{i1} = \text{Total for the first level}$
$T_{i2} = \text{Total for the second level}$

Example
For factor C
$$
\begin{aligned}
S_C &= [((C1 - C2)^2)/8 \times W_1] + [((C1 - C2)^2)/8 \times W_2] \\
&= [((3 - 0)^2)/8 \times 4.27] + [((4 - 0)^2) \times 4.0] \\
&= 4.8 + 8 \\
&= 12.8
\end{aligned}
$$

Sums of squares

When using accumulation analysis, we have already said that the calculations for the sum of squares are different. For attribute data using accumulation analysis, the total sum of squares is obtained by taking the total number of observations in the experiment, and multiplying that by the number of classes less one.

In accumulation analysis, it is very common to multiply things by the number of classes less one. This is because the final class of data is not analysed. In other words, the number of classes less one is directly comparable to the number of degrees of freedom in the more conventional ANOVA.

In this example, the total number of observations is eight and there are three classes, so the total sum of squares is 8 times 2, or 16 (Table 6.62).

The individual sums of squares for the factors and interactions are calculated in a slightly more complicated way. The sum of squares for any individual factor consists of an element due to class I, and another element due to class II.

Table 6.64   Sum of squares for error terms

Method 1

S_e = Total of the sum of squares for each error column

Method 2

S_e = Total sum of squares − Total sum of squares for known factors and interactions

In this worked example, there is one error column (G), and the first method is illustrated

$S_e = [((2 − 1)^2)/8 \times 4.27] + [((2 − 2)^2)/8 \times 4.0]$

$\quad = 0.53$

Alternatively, the calculation from the deduction of individual S from the total produces this result:

$S_e = 16 − (0.53 + 0.53 + \cdots + 0.53)$

$\quad = 16 − 15.45$

$\quad = 0.55$

The element due to a particular class is obtained by taking the sum of squares for the individual factor, and multiplying it by the weighting for that particular class. The elements for each class are added together to obtain the sum of squares for the factor (Table 6.63).

As there is only one data point for each trial, there will be only one error term, that for the primary error. This sum of squares can be calculated in two ways. It can be the total sum of squares for each of the error columns, or it can be the overall total sum of squares less the sum of squares for all the individual factors (Table 6.64).

Degrees of freedom

The total degrees of freedom are calculated by taking the total number of observations minus one, and multiplying it by the number of classes less one. So, in this case, there are eight observations and three classes, which means that the total degrees of freedom will be:

$$\text{Total degrees of freedom} = (8 − 1) * (3 − 1) = 14$$

The factor degrees of freedom are calculated by taking the number of levels for that factor, less one, and multiplying it by the number of classes minus one. In this case, there are only two levels and three classes, so the individual factor degrees of freedom will be:

$$\text{Factor degrees of freedom} = (2 − 1) * (3 − 1) = 2$$

Table 6.65 Degrees of freedom

Total
DF_{tot} = (No. of observations − 1) × (No. of classes − 1)

Factors
DF_i = (No. of levels − 1) × (No. of classes − 1)

Interactions
DF_{ij} = (No. of levels of i − 1) × (No. of levels of j − 1) × (No. of classes − 1)

The interaction degrees of freedom are obtained by taking the number of levels for each factor and subtracting one, multiplying these together, and finally multiplying the result by the number of classes less one. In this example, if there were any interactions the result would be:

Interaction degrees of freedom = (2 − 1) * (2 − 1) * (3 − 1) = 2

Table 6.65 summarizes the degrees of freedom calculations.

Variance

Although the variance, or mean square, is calculated in exactly the same way as before, there is now more than one degree of freedom associated with each factor. Consequently, the variance is no longer numerically the same as the sum of squares. The ANOVA proceeds exactly as it would for continuous data (Table 6.66).

Pooled analysis of variance

The rules for pooling terms into the error sum of squares still apply, namely, all terms with F-ratios less than one are pooled, and then the smaller factors until half the terms remain.

In this example, there are no factors with F-ratios less than one. We therefore pool sufficient factors to reduce the number to one half the original. As all are equal except one this is the only factor left.

The pooled error term is formed for both the sums of squares and the degrees of freedom. New F-ratios are calculated and the pooled ANOVA table is produced.

Inspection of the table shows that this factor (C) tests as statistically significant. It contributes substantially to the overall variance, and can

Table 6.31 Minimize viscosity totals table

Factor	Level	Total	Mean
A	1	137	6.9
A	2	174	8.7
B	1	138	6.9
B	2	173	8.7
C	1	174	8.7
C	2	137	6.9
D	1	163	8.2
D	2	148	7.4
D ∩ C	1	138	
D ∩ C	2	173	
B ∩ C	1	155	
B ∩ C	2	156	
Error	1	152	
Error	2	159	
Total		311	7.8

made in trials 2, 4, 6 and 8. These are the trials in which level 1 of factor C was used, as can be seen from the data table, column 4 (Table 6.31).

Next, the sum of squares is calculated for the total and for each of the factors.

The error term, now known as primary error, due to the vacant columns, is calculated in the same way as other factors. This is the between-experiment error.

When there are repetitions, there is a second kind of variation, the within-experiments error. Referred to as the secondary error, it is calculated by deducting the sum of all the sums of squares from the total (Table 6.32).

The formulae for the degrees of freedom are exactly the same as before. The secondary error degrees of freedom are calculated by deduction, and the primary error degrees of freedom are obtained by adding the degrees of each of the vacant columns.

The initial ANOVA is shown in Table 6.33. Before calculating the F-statistics, we have to decide how to use the error variances. If V_{e1} is less than V_{e2}, then the two terms are pooled together. If V_{e1} is greater than V_{e2} then the secondary error is eliminated from the analysis and only the primary error is used.

Table 6.66 Initial ANOVA table

Factor	DF	SS	V	F	Pool?
A	2	0.53	0.27	1	x
B	2	0.53	0.27	1	x
C	2	12.80	6.40	24.15	
D	2	0.53	0.27	1	x
E	2	0.53	0.27	1	x
F	2	0.53	0.27	1	x
Error (G)	2	0.53	0.27		
Total		16	8		

Table 6.67 Final ANOVA table

Factor	DF	SS	V	F	S'	%
C	2	12.8	6.4	23.7	12.26	77
Error	12	3.2	0.27		3.74	23
Total	14	16			16	

Table 6.68 Non-accumulating percentage table

Factor	Level	VL	SS	C	Total	VL	SS	C	Total
C	1	3	1	0	4	75	25	0	100
	2	0	0	4	4	0	0	100	100

be expected to give significant improvement when used for the process under investigation (Table 6.67).

Non-accumulating percentage table

To obtain the optimum criteria for the process, it is necessary to use a new type of table, known as the non-accumulating percentage table. This is similar to the totals table used before.

In this case only factor C is involved. If more factors had been significant, they too would be analysed in the same way. Level 1 of factor C was used in trials 1, 2, 7 and 8. Three of these yielded very little and one some smoke. The percentages are shown in the adjacent part of Table 6.68. The objective of the experiment was to minimize smoking, so level 1 of factor C is selected.

Table 6.69   Decibel values

db of %	$-10 \log(p/(1 - p))$

Table 6.70   Predicted optimum condition

For each class separately:

db of u = db of T'' + (db of $C1''$ − db of T'')

Predicting optimum conditions

Under accumulation analysis, the prediction that we need to make for our confirmatory trials is an evaluation of the percentage of product that would fall into each category as a result of the trial. With attribute data, the percentages from different factors cannot be added together in the same way as measurement contributions can. To overcome this, Taguchi proposed the use of 'decibel values' (Table 6.69).

Each percentage is translated into a decibel value, and these values can then be added together. From these the actual value can be obtained for the percentage of that class (Table 6.70). Appendix 4 gives the translation of percentages into decibel values.

For class I, the total percentage was 3 out of 8 or 37.5%. Using Appendix 4, the decibel value for 37.5% is −2.22. The total number of observations falling in class I with the first level of factor C was 3 out of 4, which is 75%. This has a decibel value of 4.77.

Inserting these values into the formula (Table 6.70), a total decibel value of u for class I is 4.77. If more than one factor had been significant, then these factors would also need to be looked up in the decibel values table. From Appendix 4, 4.77 corresponds to 75%. This predicts that 75% of the product will fall into the first class by using the selected level for the significant factor.

Exactly the same procedure is used to establish the percentage expected for class II. In this instance, the full 100% is expected within class II, which represents the sum of the very little and some smoke categories.

By deduction, the percentages falling into the very little, some smoke and charring categories can be predicted. If 75% fall in class I, then

Table 6.71 L$_8$ array with data accumulated

Trial	Factor							Grade			Class		
	A	B	C	D	E	F	G	Good	No	Bad	I	II	III
1	1	1	1	1	1	1	1	1	2	0	1	3	3
2	1	1	1	2	2	2	2	2	1	0	2	3	3
3	1	2	2	1	1	2	2	0	0	3	0	0	3
4	1	2	2	2	2	1	1	0	0	3	0	0	3
5	2	1	2	1	2	1	2	0	2	1	0	2	3
6	2	1	2	2	1	2	1	0	2	1	0	2	3
7	2	2	1	1	2	2	1	0	2	1	0	2	3
8	2	2	1	2	1	1	2	0	0	3	0	0	3
							Totals	3	9	12	3	12	24

75% fall in very little, and if all 100% are in the first two categories, then 25% must be in the some smoke bin and 0% in the charring one.

The final step is to run the process using the predicted optimum conditions, and see whether the process reproduces the predicted results.

Accumulation analysis with more than one repetition

The next example looks at an accumulation analysis where more than one observation is made for each trial. It is always a good idea to carry out more than one repetition of an experiment, where resources allow, and this is especially important in the case of accumulation analysis.

For this design, there are three grades of product: good, normal and bad. The accumulation classes are made up of:

- class I: the frequency of good product
- class II: the frequency of good plus normal product
- class III: the frequency of good, normal and bad product.

The experiment was performed using an L$_8$ array and the results with data accumulated are shown in Table 6.71.

In trial 1 one repeat was scored good and two were scored normal. The frequency of class I is therefore one, that of class II is three (the sum of one good and two normal) and the score for class III is also three, as there are no further items in the bad category.

Table 6.72 Weightings

$$W_1 = 1/((3/24) \times (1 - (3/24)))$$
$$= 9.14$$
$$W_2 = 1/((12/24) \times (1 - (3/24)))$$
$$= 4.00$$

Proportions

The next stage is to calculate the proportions for each class by dividing the number of observations in that class by the total number of observations:

$$P_I = 3/24$$

$$P_{II} = 12/24$$

Weightings

Using these proportions, weightings are then calculated for the individual classes. The weighting for a class is the reciprocal of the proportion, times one minus the proportion (Table 6.72).

Total sum of squares

The sum of squares for each factor is obtained by using the same formula as before, namely, the product of the number of observations and the number of classes less one.

$$S_T = \text{(Total number of observations)} * \text{(Number of classes} - 1)$$

$$S_T = (24) * (3 - 1) = 48$$

In this case there are 24 observations in all, and three classes. Again, the calculation is obtained by subtracting one from the number of classes, because class III is predetermined by the other classes.

Sum of squares for individual factors

The formula for calculating sums of squares for individual factors is unchanged:

$$S_i = (S_{i1} * W_I) + (S_{i2} * W_{II})$$

where $S_{i1} = (T_{i1} - T_{i2})/\text{(Total number of observations)}$

Table 6.73   Sum of squares for error terms

S_{e1} (primary) $= 0$
S_{e2} (secondary) $=$ Total SS $-$ (Sum of individual factors' sums of squares)
                     $= 48 - (27.83)$
                     $= 20.17$

$$T_{i1} = \text{Total for the first level of factor } i$$

$$T_{i2} = \text{Total for the second level of factor } i$$

Examining factor A, at level 1 for class I, there are three items scored (one from trial 1 and two from trial 2). At level 2 of factor A in class I, there are no items scored. Again for factor A, but for class II at level 1, there are six items, and at level 2 there are also six items.

The calculation of the sum of squares for factor A is:

$$S_A = ((3 - 0)^2/24 * 9.14) + ((6 - 6)^2/24 * 4.00) = 3.43$$

Sum of squares for error terms

As there is more than one repetition in this example, there will be two error terms: a primary and a secondary. Remember, the primary error term is the between-experiment error and the secondary is the within-experiment error. In this example, however, there are no vacant columns, so the primary error will be zero.

The easiest method of calculating the secondary error term is almost always by deduction. The total sum of squares is known, and from that the sum of all the individual sums of squares can be taken (Table 6.73).

Degrees of freedom

The total degrees of freedom is obtained by subtracting one from the number of observations, and one from the number of classes, and then multiplying the two figures together.

$$DF(\text{total}) = (24 - 1) * (3 - 1) = 23 * 2 = 46$$

The degrees of freedom for each factor is calculated by the number of levels less one, times the number of classes less one:

$$DF(\text{factor}) = (2 - 1) * (3 - 1) = 1 * 2 = 2$$

Since there are no vacant columns, the degrees of freedom for the primary error will be zero. The degrees of freedom for the secondary error term are obtained through deduction. In this case there are 46 total degrees of freedom, and there are seven factors, each with two degrees of freedom.

Thus, there will be $46 - (7 * 2)$ or 32 degrees of freedom associated with the secondary error term.

Pooled error terms

As the variance of the primary error is less than the variance of the secondary error, the sums of squares and degrees of freedom for the two terms are pooled by addition.

$$S_{e(\text{pooled})} = S_{e1} + S_{e2} = 0 + 20.17 = 20.17$$

$$DF_{e(\text{pooled})} = DF_{e1} + DF_{e2} = 0 + 32 = 32$$

$$V_{e(\text{pooled})} = S_e/DF_e = 20.17/32 = 0.63$$

Initial ANOVA

Table 6.74 shows the initial ANOVA. Using the usual pooling conditions, all factors with F-ratios of less than one are pooled into the error

Table 6.74 Initial ANOVA table

Factor	DF	SS	V	F	Pool?
A	2	3.43	1.7	2.7	
B	2	14.10	7.1	11.2	
C	2	6.10	3.1	4.8	
D	2	1.05	0.5	0.8	x
E	2	1.05	0.5	0.8	x
F	2	1.05	0.5	0.8	x
G	2	1.05	0.5	0.8	x
Error	32	20.17	0.6		
Total	46	48.00			

term. This means that factors D, E, F and G are pooled. This leaves three out of seven factors, and so the second condition of pooling is not needed.

The final ANOVA table retains factors A, B and C (Table 6.75). Only factor B is significant when tested against the F-statistics. The error term still accounts for more than 50% of the variation, which is not very satisfactory. We shall complete the analysis of this example, but it would be worthwhile considering additional experiments to improve our understanding of the problem. In this case, no interaction terms were incorporated. An L_{16} array allowing additional factors and interaction terms may be useful.

Selection of levels

Table 6.76 shows the non-accumulating percentage.

The final stages are the selection and prediction of optimum criteria and expected conditions. For factor B at level 1 there are three good, seven normal and two bad items. At level 2 there are no good, two normal and ten bad items. It therefore seems sensible to accept level 1 as the optimum condition to use.

For each class individually, the predicted optimum condition is calculated (Table 6.77).

Table 6.75 Final ANOVA table

Factor	DF	SS	V	F	S'	%
A	2	3.43	1.72	2.81	2.21	4.6
B	2	14.10	7.05	11.57	12.88	26.8
C	2	6.10	3.05	5.01	4.88	10.2
Error	40	24.37	0.61		28.03	58.4
Total	46	48.00				

Table 6.76 Non-accumulating percentage table

Effect	G	N	B	Total	G	N	B	Total %
B1	3	7	2	12	25	58	17	100
B2	0	2	10	12	0	17	83	100

Table 6.77   Predicted optimum conditions

For class I
db of u = db of T'' + (db of B'' − db of T'')
where
$$T'' = 3/24 = 12.5\%$$
$$B'' = 3/12 = 25.0\%$$
db of u = −8.4 + (−4.77 − (−8.4))
        = −4.77

From Appendix 1
   $P = 25\%$

For class II
where
$$T'' = 12/24 = 50\%$$
$$B'' = 10/12 = 42\%$$
db of u = 0 + (−1.4 − 0)
        = −1.4
$$P = 42\%$$

Interpretation

For this example, there is a predicted percentage of 25% that will fall into class I, and a predicted percentage of 42% that will fall into class II. Translating to the original categories: 25% will fall into the good bin, 42 minus 25 (17%) into the normal bin and the remaining 58% into the bad group.

The relevant levels would be inserted and the confirmation process run to verify these conclusions.

Signal-to-noise ratio

So far we have discussed experiments that are geared towards centring, or targeting, processes. Unlike most other experimental design methods, Taguchi's techniques also allow us to study the variation of a process, and ultimately to optimize the process for variability, as well as target. This is what Taguchi's concept of signal-to-noise ratios is concerned with.

Signal-to-noise ratios are used in several ways. One of these is the analysis of noise factors. These are the types of factor that are known to

Table 6.78 The need for signal-to-noise ratios

Trial	A	B	C	Outcomes			Total
1	1	1	1	19	20	21	60
2	1	2	2	15	20	25	60
3	2	1	2	19	20	21	60
4	2	2	1	15	20	25	60

The difference between the totals for each level is zero, so the sums of squares for each factor is also zero:

$$S_A = ((\Sigma A1 - \Sigma A2)^2)/N$$
$$= ((120 - 120)^2)/12$$
$$= 0$$

But there is a difference between the trials. Normally, sum of squares measures the variability of the mean; here, though, the means are the same. The difference is in the variability within an experimental run

cause variation, but that cannot normally be controlled. Examples of this might be seasonal variation, sources of raw materials, different machines or different product lines in a plant.

Taguchi advocated including noise factors rather than excluding them; this means that results are highly reproducible. Traditional experimental techniques block them out, which means that variability cannot be studied.

Many people trying to resolve problems do so by eliminating factors. Unfortunately, this rarely succeeds, and when it does the factors identified can rarely be avoided, so nothing is gained. The Japanese approach is to minimize the effect of these noise factors, without embarking on a witch-hunt.

Signal-to-noise ratios are used in two ways: first, to minimize the variation induced by repetitions and, secondly, to analyse variation deliberately induced by noise factors.

Signal-to-noise analysis for repetitions

We will begin by examining the effect of repetitions. Consider the L_4 array in Table 6.78. The signal-to-noise ratio is calculated for each experimental trial in a design. The formula changes depending on the purpose of the experiment, to minimize or to maximize (Table 6.79). With the two formulae, regardless of the purpose of the experiment, the

Table 6.79   Signal-to-noise ratio

1. -10 eliminates decimals
2. Log neutralizes extreme values
3. Sum of squares, but without subtracting the correction factor, $\Sigma(Y^2)/2$

Calculation when maximizing a quality characteristic
S:N max $= -10 \log[1/n \times (1/(y1^2) + 1/(y2^2) + \cdots + 1/(yn^2))]$

Calculation when minimizing a quality characteristic
S:N min $= -10 \log[1/n \times ((y1^2) + (y2^2) + \cdots + (yn^2))]$

Table 6.80   Signal-to-noise ratios

Trial	A	B	C	Outcomes			Total	S:N
1	1	1	1	19	20	21	60	26.00
2	1	2	2	15	20	25	60	25.45
3	2	1	2	19	20	21	60	26.00
4	2	2	1	15	20	25	60	25.45

For trials 1 and 3

S:N max $= -10 \log[1/3 \times (1/(19^2)) + (1/(20^2)) + (1/(21^2))]$
$\qquad\quad -10 \log[0.0102]$
$\qquad\qquad 26$

For trials 2 and 4

S:N max $= -10 \log[1/3 \times (1/(15^2)) + (1/(20^2)) + (1/(25^2))]$
$\qquad\quad -10 \log[0.002848]$
$\qquad\qquad 25.45$

bigger the value the better the result. Consider the example given earlier (Table 6.80).

The analysis of signal-to-noise ratios is exactly the same as that for mean values. The first stage is to calculate the sum of squares (Table 6.81).

The signal-to-noise ratio is an approximation of the inverse of the coefficient of variation. This would represent the mean divided by the standard deviation. Bear in mind: the signal-to-noise ratio evaluates both the mean and variation of a process together. We can see this by looking at three examples (Table 6.82).

Compare groups 1 and 2. Group 1 has the same mean as group 2, but a smaller variance. The signal-to-noise ratio is larger for group 1 than group 2, showing that it has the better variance.

Table 6.81   Sum of squares for signal-to-noise ratios

$$S_A = \frac{((\Sigma A1 - \Sigma A2)^2)}{N}$$

$$= \frac{((51.45 - 51.45)^2)}{4}$$

$$= 0.0$$

$$S_B = \frac{((\Sigma B1 - \Sigma B2)^2)}{N}$$

$$= \frac{((52 - 50.9)^2)}{4}$$

$$= 0.3$$

$$S_C = \frac{((\Sigma C1 - \Sigma C2)^2)}{N}$$

$$= \frac{((51.45 - 51.45)^2)}{4}$$

$$= 0.0$$

N is the number of signal-to-noise ratios, i.e. the number of trials, not the number of repetitions

For A and C, there is no difference between the variability at level 1 and at level 2

For B, level 1 has more variability than level 2

Compare groups 2 and 3. Group 2 has the same variance as 3, but a lower mean value. The object of the experiment is to maximize the target value. The signal-to-noise ratio therefore shows that group 3 is the better, as it has a higher mean, although it has the same variance.

The same comparison process can be used when the object is to minimize the target value (Table 6.83).

Compare group 1 with group 3. The former has the smaller mean and standard deviation. This makes it the best group of all, and consequently it has the larger signal-to-noise ratio.

Signal-to-noise analysis with noise-induced repetitions

The other way in which signal-to-noise analyses are used is when noise factors are included in experiments. Noise factors can be inserted to the right of the original orthogonal array. The orthogonal array contains factors that can be controlled in the process. These are control factors,

Table 6.82   Signal-to-noise ratio (maximum) measures both mean and variability

---

Group 1
Values   = 5.5, 6.0, 6.5
Mean   = 6.0
S_D = 0.5
S:N max = $-10 \log[1/3 \times ((1/(5.5^2)) + (1/(6.0^2)) + (1/(6.5^2)))]$
        = 15.5

Group 2
Values   = 4.0, 6.0, 8.0
Mean   = 6.0
S_D = 2.0
S:N max = $-10 \log[1/3 \times ((1/(4.0^2)) + (1/(6.0^2)) + (1/(8.0^2)))]$
        = 14.5

Group 3
Values   = 5.0, 7.0, 9.0
Mean   = 7.0
S_D = 2.0
S:N max = $-10 \log[1/3 \times ((1/(5.0^2)) + (1/(7.0^2)) + (1/(9.0^2)))]$
        = 16.2

Table 6.83   Signal-to-noise ratio (minimum) measures both mean and variability

---

Group 1
Values   = 5.5, 6.0, 6.5
Mean   = 6.0
S_D = 0.5
S:N min = $-10 \log[(1/3 \times ((5.5^2) + (6.0^2) + (6.5^2)))]$
        = -15.6

Group 2
Values   = 4.0, 6.0, 8.0
Mean   = 6.0
S_D = 2.0
S:N min = $-10 \log[(1/3 \times ((4.0^2) + (6.0^2) + (8.0^2)))]$
        = -15.8

Group 3
Values   = 5.0, 7.0, 9.0
Mean   = 7.0
S_D = 2.0
S:N min = $-10 \log[(1/3 \times ((5.0^2) + (7.0^2) + (9.0^2)))]$
        = -17.1

Table 6.84   Signal-to-noise ratio

Example
Add one noise factor with two levels
Half of the replications are carried out at one level and half at the
second level

N1		N2				Total
1	3	2	7	7	3	23
9	9	9	6	5	5	43
8	8	6	5	7	4	38
Tot N1 = 55		Tot N2 = 49				

An additional sum of squares can be calculated for the noise factor:

$S_N = ((\Sigma N1 - \Sigma N2)^2)/\text{Total } N$

This difference, which represents variation across experiments, is
not relevant here. However, it must be removed from the error term.

Table 6.85   Noise factors

Trial	Factor			DOG 1			DOG 2			Totals	S:N min
	A	B	C								
1	1	1	1	4	6	4	11	10	12	47	−18.58
2	1	2	2	7	8	7	8	6	8	47	−17.35
3	2	1	2	7	7	6	9	9	9	47	−17.98
4	2	2	1	3	4	5	13	12	10	47	−18.87

and they are defined as those factors that impart a significant effect
on the quality characteristic. Noise factors may be inserted to the right
of the array, as they cannot normally be controlled in the working
process. The reason for including them is to find the control factors that
minimize the variation while the noise factor is working (Table 6.84).

A simple example can be used to illustrate the use of this form
of analysis (Table 6.85). The totals are all the same, so an analysis of
mean value would not be of any benefit, but if you were to calculate
the sum of squares for the noise factor, you would find that it is
very high. We shall look at one example in which noise factors were
used.

Case study: signal loss using infrared data transmission

This experiment was carried out by a computer magazine to test the flexibility of infrared (IRDA) networking. The objective of the experiment was to minimize the signal loss from an interoffice infrared data transmitter.

Four factors were studied, each at two levels, and three interactions were included. The noise factor consisted of daylight entering the office suite. This is outside normal control, but could be selected when running experiments.

An L_8 array was used, and this is shown in Tables 6.86–6.89, along with the results of the six replicates, the signal-to-noise ratios and the tables from the ANOVAs.

Table 6.86   Case study: infrared transmission

	Factor	Levels chosen	
A	Lens angle	64	102
B	Transmitter supplier	Data gap	Univ. force
C	Receiver type	Data gap	Univ. force
D	Transmission speed	2400	115 200

Interactions

B ∩ C
B ∩ D
C ∩ D

Noise factor

Daylight strength	Bright	Dull

Table 6.87   Case study: infrared transmission

Trial	Factor								Noise factor						S:N
	A	B	B ∩ D	C	C ∩ D		B ∩ C	D	Bright			Dull			
1	1	1	1	1	1		1	1	0.4	0.4	0.3	0.4	0.4	0.5	7.9
2	1	1	1	2	2		2	2	0.1	0.1	0.2	0.3	0.4	0.3	11.8
3	1	2	2	1	1		2	2	0.1	0.1	0.1	0.2	0.1	0.2	16.9
4	1	2	2	2	2		1	1	0.1	0.4	0.3	0.4	0.6	0.4	8.1
5	2	1	2	1	2		1	2	0	0	0.1	0.3	1.3	0.1	5.2
6	2	1	2	2	1		2	1	0.1	0.1	0.1	0.1	0.2	0.3	15.5
7	2	2	1	1	2		2	1	0.1	0	0.2	0.3	0.4	0.3	11.9
8	2	2	1	2	1		1	2	0.1	0.2	0.2	1.5	0.6	0.2	3.4

From the mean value analysis, the error percentage contribution is found to be particularly high. Although two factors represent much of the variation and the noise characteristic is the largest of these, at 19%, a considerable proportion of the variance is still not explained.

The signal-to-noise analysis shows a percentage contribution for the error, or unassignable variation, of 13%. This indicates significant results when both the mean value and the variation are included. The most significant variable was receiver type, which accounted for 76% of the variance.

By deduction, as receiver type was not significant for mean value alone, but was for the signal-to-noise analysis, this was a significant factor controlling variation (Tables 6.90 and 6.91).

It can be seen that there is a significant interaction between the receiver type and the angle of the transmission lens. One maker's receiver worked better than the other across a wider range of beam dispersions.

Table 6.88 Infrared data transmission: initial ANOVA table (mean)

Source	DF	S	V	F
B	1	0.00	0.00	0.00
B ∩ C	1	0.01	0.01	0.17
C ∩ D	1	0.10	0.10	1.67
B ∩ D	1	0.02	0.02	0.33
D	1	0.00	0.00	0.00
C	1	0.48	0.48	8.03
A	1	0.00	0.00	0.00
Noise	1	0.75	0.75	12.55
Error	39	2.33	0.06	

Table 6.89 Infrared data transmission: final ANOVA table (mean)

Source	DF	S	V	F	S'	P%
C ∩ D	1	0.10	0.10	1.86	0.05	1
C	1	0.48	0.48	8.95	0.43	12
Noise	1	0.75	0.75	13.98	0.70	19
Error	44	2.36	0.05		2.51	68

Table 6.90 Infrared transmission: initial ANOVA table (signal-to-noise ratio)

Source	DF	S	V
B	1	9.43	9.43
B ∩ C	1	0.00	0.00
C ∩ D	1	14.65	14.65
B ∩ D	1	1.33	1.33
D	1	5.81	5.81
C	1	124.26	124.26
A	1	4.33	4.33
Total	7	159.81	

Table 6.91 Infrared transmission: final ANOVA table (signal-to-noise ratio)

Source	DF	S	V	F	S'	P%
B	1	9.43	9.43	3.28	6.6	4
C ∩ D	1	124.30	124.30	43.23	121.4	76
C	1	14.70	14.70	5.11	11.8	7
Error	4	11.50	2.88			13
Total	7	159.81	22.83			100

Signal-to-noise analysis with many noise factors

The final topic to cover in this signal-to-noise section is the use of more than one noise factor in a design. If more than one noise factor is identified for an experiment, then an outer array is formed of the noise factors in exactly the same way as the inner array would be of the control factors (Figure 6.7).

This outer array is constructed in exactly the same way as the inner, and the size of the array is determined by the number of noise factors to be incorporated. The number of rows of this array determines the minimum number of replicates that can be used for the experiment.

Table 6.92 shows an example using an L_8 inner array (up to seven control factors) with an L_4 outer array (three noise factors).

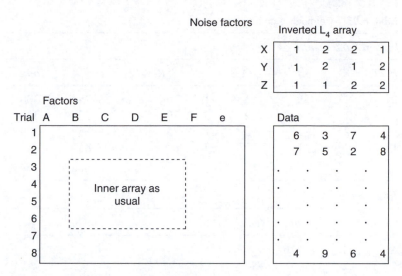

Figure 6.7 Signal-to-noise and multiple noise factors

Table 6.92 Calculating confidence limits

Confidence interval (CI) for the predicted value of the confirmation condition
   $F95$ = F-statistic from tables using 1,DF err degrees of freedom

N(effective) = Effective number of replications

$$= \frac{\text{No. of trials}}{\text{DF of sources considered for prediction}}$$

$$\text{CI} = \sqrt{\frac{(F95 \times V(\text{error}))}{N(\text{effective})}}$$

Confidence interval (CI) for the confirmation run result
   r = No. of runs in the confirmation trial

$$\text{CI} = \sqrt{[F95 \times V(\text{error}) \times (1/N(\text{effective}) + 1/r)]}$$

Confidence limits

Following a Taguchi-designed experiment, we have seen that it is possible to calculate a predicted optimum condition. The process need not centre on this value; it is a prediction that should be used to confirm that the experimental analysis was successful.

Table 6.93 Confidence limits

L_4 array and results

Trial	A	B	$A \cap B$		Results			Total
1	1	1	1	71	76	78	75	300
2	1	2	2	83	79	75	79	316
3	2	1	2	81	75	77	75	308
4	2	2	1	87	87	85	81	340
							Total	1264

Totals table

Factor	Level	Total	Mean
A	1	616	77
A	2	648	81
B	1	608	76
B	2	656	82

Initial ANOVA

Source	DF	SS	V	F
A	1	64	64.0	7.25
B	1	144	144.0	16.30
$A \cap B$	1	16	16.0	1.81
e2	12	106	8.8	
Total	15	330		

$F(95\%, 1,12) = 4.75$
$F(99\%, 1,12) = 9.33$

Final ANOVA

Source	DF	SS	V	F	SS'	%
A	1	64	64.0	6.82	54.6	16.5
B	1	144	144.0	15.34	134.6	40.8
e2	13	122	9.4		140.8	42.7
Total	15	330			330.0	100

Table 6.94 Confidence limits

$U = (A2' - T') + (B2' - T') + T'$
$= 81 + 82 - 79$
$= 84$

Confidence limit for the predicted response

$$CI = \sqrt{\frac{(F95 \times V(\text{error}))}{N(\text{effective})}}$$

$$= \sqrt{\left(\frac{4.67 \times 9.4}{(16/(1+1+1))}\right)}$$

$= 2.9$

Confidence interval for the confirmation run result

$$CI = \sqrt{[F95 \times V(\text{error}) \times (1/N(\text{effective}) + 1/r)]}$$

$$= \sqrt{[4.67 \times 9.4 \times (1/\{16/(1+1+1)\} + 1/4)]}$$

$= 4.8$

Confidence interval for predicted values

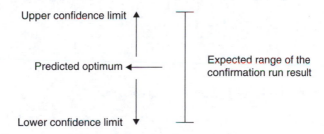

Confidence interval for confirmation run

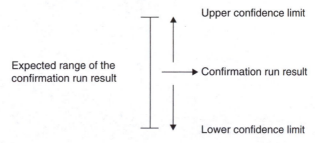

Figure 6.8 Confidence limits

The confirmation run generates a result that should be compared with the prediction. If the two coincide, then the experimental analysis has clearly been successful. To measure their closeness confidence limits can be calculated, either for the prediction or for the confirmation run (Tables 6.93 and 6.94).

In the case of the prediction, these limits represent the range within which the true value should lie on 19 out of 20 occasions. For the confirmation run result, the limits represent the range within which the predicted value should fall on 19 out of 20 occasions. These situations are represented in Figure 6.8.

References

Box, G. E. and Draper, N. R. (1969) *Evolutionary Operations*. London: John Wiley.

Taguchi, G. (1968) *Introduction to Quality Engineering*. Tokyo: Asian Productivity Organization.

Wilson, G. B., Bendell, A. and Millar, R. E. M. (1990) *Taguchi Methodology Within Total Quality*. Kempston, Bedford: IFS International.

7

Statistical process control

What is statistical process control?

In its modern form, statistical process control (SPC) has been available for nearly 70 years. It is a form of charting, based on sound statistical principles, which allows us to see how product or service processes are performing. Not only can SPC tell us when something is going wrong or beginning to go wrong, but when alarming changes occur, it can also tell us whether there is any point in trying to do something to remedy them.

Properly applied, SPC is virtually foolproof. It is simple to use, involves little or no complicated mathematics, and almost guarantees to pay for itself in saved effort. However, despite such a long pedigree and all its selling features, SPC probably has one of the poorest adoption records of any management technique.

There are quite a number of reasons for this, although I suspect that the most significant is the first word in its title: statistical. If you do not beat them to it, almost every SPC course has some wit who will recall Disraeli's comment: 'There are lies, damned lies, and statistics'. For people who have a perception that they are not that good at mathematics, the word strikes a fear that they will not be able to cope. Most schools opt for the bizarrely misnamed 'applied mathematics', rather than statistics, as an A-level option. Statistics is sometimes seen as an irrelevance, while to others it is a 'soft' science, almost an art. Even university students have been known to look down on their statistics options as lacking rigour and worth.

In practice, statistics probably has more industrial applications than any form of applied mathematics (of course, it is applied mathematics, but that phrase is strangely reserved for mechanics). It is perfectly rigorous, but deals with much larger numbers of events at any one time. This gives it a 'cloud-like' quality, which some people mistake for lack of clarity. In many cases, because it deals with the real world, the use of statistics is no harder than drawing a simple graph, unlike the complex proofs of mechanics.

As for that quote by Disraeli, he was not even talking of the same kind of statistics that we are here. His were the kind of facts that politicians often deal in: simple statements of numbers, like the number of unemployed or the number of trains running late on a particular line.

Where SPC is found, it is usually in the manufacturing plant, and then often only on the production line itself. For some reason, people find it much easier to identify measurable features of a manufacturing process than a service one. This could not be further from the truth. We mentioned the Japanese holiday centre in Chapter 4. The following list of potential applications was generated during a short, warm-up, brainstorming session by a small group of people from the customer services department of a computer software company (Table 7.1).

Although we may argue with the rationale behind a few of these points, there are circumstances where each and every one might be appropriate. For example, using SPC to monitor the number of customer complaints may imply that there is an acceptable level of complaints. If you seriously believe this, then six sigma is going to remain out of your grasp for the time being. As the IBM advertisement in the late 1970s said: 'If your defect rate is one in a million – what do you tell that customer?'

Statistical process control can be successfully applied in all sorts of situations. For example, I have often worked in consultancies, where much of the marketing effort revolves around media relations. On average, we generate between five and ten leads from each editorial mention. We aim to generate twenty new leads this way each month, so we have a corresponding target of between four and five items of PR. Enquiries may come directly to us or via the magazine concerned.

Our marketing co-ordinator monitors the number of enquiries received, and the results are plotted. She knows that the range of enquiries can be between five and twenty-eight. Fewer than that would suggest a shortfall of leads, requiring investigation; more than that and we need to take

Table 7.1   An ad hoc list of possible service applications of statistical process control

1. Difference between the original estimate and actual time to fix an item
2. Number of appraisals carried out in a week
3. Number of errors found in project reports
4. Hours of unplanned overtime
5. Number of corrections to job vouchers
6. Number of crisis meetings held
7. Number of calls to the help desk
8. Cost of technical support provided to the sales department
9. Number of invoices sent late or with errors
10. Amount of photocopier downtime
11. Number of customer complaints
12. Time taken to answer the telephone
13. Number of new customer contacts made
14. Frequency of office/staff reorganizations
15. Number of sales bids won
16. Age of debts
17. Profit, sales and costs
18. Time spent managing by wandering around
19. Online service availability
20. Staff turnover
21. Amount of work in progress

some action to make sure that we can handle them all satisfactorily. This is not quite the full story as there are also 'runs tests', which she uses to catch incipient problems. In a professional consultancy practice, just like most other services, there is a wide range of possible applications for SPC. Among others that we encourage our consultants to chart are their time utilization and their sales figures.

A typical individual's chart is shown in Figure 7.1. In this case, it is for a depot of a construction company. It is amazing how often a manager is appointed to a position with a clear brief to achieve some kind of change and yet does not collect information to show that they have done what was expected. In this particular case, a new manager was appointed to a depot of a construction company. His brief was to do something about morale. He chose two measures of this: stock losses and absenteeism. The chart shown is the one produced for absenteeism.

The individual's chart is just one of a number of different control charts. We shall look at the most useful ones in detail later in this chapter.

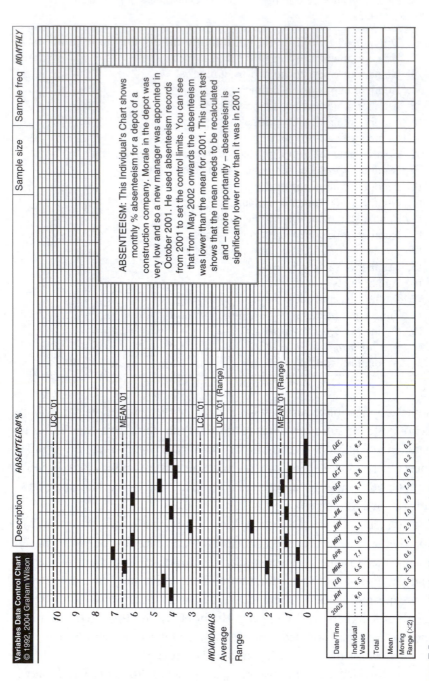

Figure 7.1 An individual's chart for absenteeism

The objective of using SPC, then, is to monitor factors that you know are important, so that you can detect when they are changing, and in particular to determine whether those changes are worth worrying about or simply natural variation. Unfortunately, in the past SPC has been applied as a simple tool without much concern for its relevance to a given set of circumstances and, far more often than not, by the managers overseeing an operation rather than by the operators themselves.

What is different now?

En route to six sigma there are some important distinctions in the way in which SPC will be applied. There is the culture, which you should already be well on the way towards changing. Teams should be working under their own supervision, with 'managers' responsible for teaching and developing new skills. Through quality function deployment (QFD) you have tied customers' expectations into your organization's technical performance, shedding irrelevant or superfluous activities. The causes of variation in the key technical factors are in the process of being isolated and quantified using Taguchi's techniques. The improvement process should be well ahead of your competitors.

The number of companies that have progressed this far could probably be counted on the fingers of one hand. Statistical process control is the final stage. We have already removed most of the assignable variation. The role of SPC is to see that it stays that way.

Is it right to 'control'?

I have pointed out already the problems of control as a non-value adding activity, arguing that most inspection activities are redundant. When SPC is applied in the traditional sense members of the quality department typically do it. Lots of quality managers like to deny this, but only a month ago, I was in the office of just such a person, from a massive German manufacturing company. Only seconds after he had told me how the people in production did all the quality control, one of his colleagues (although, from the way they interacted, it was clearly a boss–subordinate relationship) came into the room. She sat at her terminal, typed in some numbers that she had just collected from the production area and printed out some control charts. Not a lot of empowerment there, I suspect, and, from what they said, quite a long way to go before they can dream of six sigma.

Even today, in many companies, SPC is used on samples taken from the end of the production process, or at least from partially completed stages.

One textile manufacturer which I was asked to help insisted that they had one of the most sophisticated production processes in their industry. It had electrical controls all over the place, and these fed back to a master control room from which the dozen or so production lines could all be seen and controlled. The quality inspectors had been progressively creamed off from the production supervisors; any good supervisor was offered the chance to become an inspector, which was seen as the next step in the management chain.

Samples were taken of the product as it emerged from the process. Each was tested for a number of mechanical parameters and then assessed for colour. Despite all this sophistication and control, the company still had production problems. Between 10 and 40 per cent of its products had to be downgraded and sold at a lower profit.

It was only when the inspectors were transferred back onto the line and given responsibility for monitoring in-process variables, rather than product ones, that these defects could be eradicated. Their control was now exerted over bath temperatures, transit speeds, alkali concentrations and periods of washing. This was only after a long period spent re-evaluating the customers' needs, redesigning the production plant layout and carrying out a series of designed experiments to establish the key process variables.

Product control

The traditional approach that the textile producer was following involved product control. Most product control activities allow wastage to occur. Statistical process control that is applied to products, even if they are only partially processed, is not likely to be effective. It will certainly not be as efficient and is probably concealing losses. Product control is a detection strategy. In modern manufacturing practice, product control is no longer justified except in very rare circumstances (and do not fool yourself, you are not that special).

One further example of the remarkable inadequacy of detection strategies comes from a UK car-maker in the early 1980s. The story, which is probably apocryphal, says that its hatchback production line spent most of one weekend producing four-door hatchbacks. (Think about it!) The beauty of apocryphal tales is most of us actually think that they might be true.

It was the product control practices of the 1960s that led to a vast bureaucracy of quality assurance (QA) created in many businesses. This was simply exacerbated by the era of so-called quality management systems in the 1980s and early 1990s. At one point, the DTI was actively promoting the ISO9000 series of standards for quality management systems, while the Cabinet Office were publishing charts that showed that productivity exponentially declined with the number of quality systems implemented.

Process control

Product control can only happen after the event. By contrast, a prevention strategy would be based on control of the production process itself (including the raw materials going into it). It is much more effective and, in the long run, much more efficient.

Since 2001 performance targets have been set by the government for Ambulance Trusts. They say that for life-threatening 999 calls, 75 per cent must result in a Trust-dispatched response arriving within eight minutes, and in 95 per cent of cases a fully crewed ambulance must arrive within nineteen minutes. By 'response', they allow for a fully crewed ambulance, a first response vehicle (typically a car or motorbike with a paramedic) or a community first responder trained by the ambulance service. Process control of this kind has undoubtedly resulted in some exceptional initiatives and a dramatically different approach to responding in many parts of the country.

Ironically, though, the choice of eight and nineteen minutes is an unfortunate one, as it prevents equal time intervals to be used for statistical analysis, thereby making it quite unlikely that Trusts will use sophisticated monitoring of the data. Had 9 and 18, or 8 and 18, or 8 and 20 been used the analysis would be a lot simpler. Bear this in mind when you devise your own standards.

There is a further flaw in this kind of thinking that is highlighted later. It could be argued that a maximum time should also be specified; for example, after a 999 call there should never be a wait of longer than, say, 60 minutes. No matter how difficult this might be to guarantee, it would mean that exceptions could be reported and genuine problem solving be put into place.

Ten years ago, the post office operated product control on the emptying of pillar boxes, based on random samples of test mail. Today they have moved to process control with the collectors, in this case, using remote terminals to log the time of collection from the box.

The benefit of process control is that it gives immediate feedback on key factors that influence the product. It is not applied unless the factors have already been shown to be significant. Thus, we have moved further along the problem-solving chain, so that we can react quickly if a problem is discovered. With product control we still have to trace the problem back to its source factors. This can often take a long time, and the disruption caused can be devastating.

Two incidents in a retirement home show just how easy it is to use product control rather than process control. Twice in recent months, elderly residents suffering from progressive senile dementia have got up in the middle of the evening and walked out of the home. They both collapsed and died. The product control solution proposed by the director of the home was to introduce electronic tagging of patients. This meant that if any did go 'out of control' they would trigger an alarm as they wandered through the door. The process control solution favoured by the Health Authority and the Coroner was to improve staff recruitment, training and shift planning, and to provide more activities for the residents in the evenings, thereby keeping them occupied. As the Health Authority pointed out, the latter solution would cost very little more and would have many spin-off advantages, while the tagging would just cost money.

Are we capable of doing this?

Armed with knowledge of the technical responses that need to be optimized, and having identified the control factors that influence these, we are ready to move into the final stage of the six sigma process. In a traditional setting it is often tempting to jump the first step in this final stage.

With a firm order and an approaching deadline it is easy to assume that we already know the answer. After all, we are the best widget supplier in town, so naturally we can do it!

A definition of capability

Capability is a statistical term that measures how closely our actual performance matches the target which we are trying to achieve. Imagine that you are in charge of the Ambulance Trust dispatch centre, responsible for getting help to the scene of a reported patient within eight minutes. You would say that you were capable if your process of responding meant that you could meet the specification. If, because of lack of resources, the geographical shape of your patch or one of a hundred other causes, you found that you could not meet the specification, then you would describe yourself as not being capable.

The terminology of capability dates from a time when industry was dominated by manufacturing, so it speaks of machines. Today, the service sector has apparently overtaken the manufacturing one. Had this been the case as capability theory was unfolding, we would probably have spoken of single events and multiple events and the capability of each. Instead, we still speak of a 'machine', when we can apply exactly the same thinking to a service 'event'.

When we are dealing with a single machine we talk of 'machine capability'. Where a process consists of several different possible sources of variation, including the equipment, the people, the materials, the methods and so on, we talk of 'process capability'. The statisticians have allowed us a degree of tolerance in their mathematical treatment of the two different situations. Maintaining a capable condition in a single machine is easier than for several machines, where a problem with one machine can be compounded as the process passes through others.

The requirement for a machine to be capable, in the statistical sense, is that its average output plus or minus four standard deviations should fall within the specification. In other words, 99.994 per cent of the items produced by the machine will be within the specifications. For a process, the requirement is less demanding. Process capability is defined as 99.73 per cent of the output falling within the specifications.

Returning to the ambulance example; the specification is that a first response should arrive between zero and eight minutes of the 999 call being logged. Is the Trust meeting the requirement? Table 7.2 shows a sample of times taken by units to arrive at incidents. The data are routinely recorded anyway, so no extra effort is required.

The sample was taken at random from records representing a period of three months. Why make so much of this? It is vital that the sample should be representative of the whole. If there are seasonal variations they need to be acknowledged, and either they need to be incorporated by taking a large enough sample, or the capability needs to be reassessed in smaller chunks so that anomalies can be tested.

How to calculate sample statistics

So that we can describe the data we need to record them and calculate simple descriptive statistics, such as a measure of central tendency and of spread. In some situations it is possible to measure every item, but the time and cost are often prohibitive, so we either use samples or categorize the information. Nowadays, the processing power of PCs is such

Table 7.2 Ambulance response times

Upper time interval	No. of incidents	% of incidents	Cumulative frequency	Cumulative % frequency
2	3	6	3	6
4	6	12	9	18
6	10	20	19	38
8	12	24	31	62
10	7	14	38	76
12	3	6	41	82
14	1	2	42	84
16	2	4	44	88
18	1	2	45	90
20	0	0	45	90
22	1	2	46	92
24	1	2	47	94
26	1	2	48	96
28	1	2	49	98
30	1	2	50	100
Total	50			

that people are often tempted to try to use ridiculously large amounts of information. A sample can often be far easier to understand and interpret.

Obtaining sample statistics: where there is one sample

In the example with the ambulance arrival data, there is just one sample. The data were taken at random from the service log. This is a classic situation where someone could be tempted to try to process everything, but actually a sample is just as good.

Provided that there are sufficient data, the standard deviation of the population can be calculated according to the formula given in Chapter 2. (The note below demonstrates how the sample begins to approximate towards the population.)

Obtaining sample statistics: where there are several samples

Table 7.3 has two formulae to quickly calculate an estimated population standard deviation based on some samples.

Table 7.3 Estimating standard deviations (SD)

SD (est.) = (Average sample range)/d2
       where d2 is a constant from the following table
SD (est.) = (Average sample SD)/c4
       where c4 is a constant from the following table

Sample size, n	d2	c4
2	1.13	0.798
3	1.69	0.886
4	2.06	0.921
5	2.33	0.940
6	2.53	0.952
7	2.70	0.959
8	2.85	0.965
9	2.97	0.969
10	3.08	0.973

If we have taken a number of successive samples we can calculate the range of values in each and then take the average of these, in other words the average sample range. Dividing by the constant d2 shown in Table 7.3 gives an estimate of the population standard deviation.

Alternatively, if we have calculated the standard deviation of each of the samples (which can often be obtained using a spreadsheet), then we can go on to obtain the average of these, and by dividing this by the value of c4 shown in Table 7.3, we can again estimate the population standard deviation. As the sample size increases from ten towards twenty, the value of the constant, c4, tends towards a value of 1.000. In other words, as the sample sizes become larger their standard deviation becomes closer and closer to that of the population as a whole. Pretty obvious really!

Process capability

In most working environments, the acceptable process capability, C_p, will be just over 1.0. As noted earlier, this will ensure that three standard deviations either side of the mean will fall within the specification limits. This is fine for samples that are clearly centred between the upper and lower specification limits (USL and LSL, respectively). But if the mean changes, it could push the distribution outside the limits, and the process would be said to be out of control. C_p is therefore effectively a measure of the spread of the process, but depends on the location of the mean as to whether it is acceptable or not.

To avoid this problem a second index, C_{pk}, is used. This is a measure of both the spread and the process setting, in other words the central tendency of the process. C_{pk} is defined as the minimum of:

$$\frac{(\text{USL} - \text{mean of the sample means})}{(3 \times \text{estimated SD})}$$

or

$$\frac{(\text{Mean of the sample means} - \text{LSL})}{(3 \times \text{estimated SD})}$$

If C_{pk} is less than 1, then the process is not capable of meeting the customer's specification without making some changes.

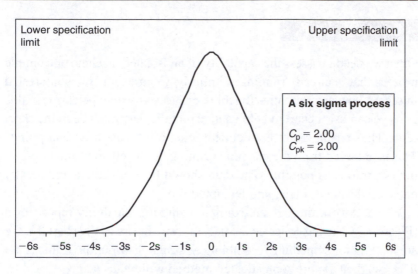

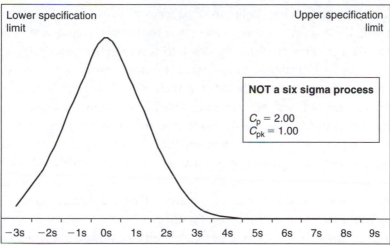

Figure 7.2 Optimizing process capability

Similar formulae apply to machine capabilities, C_m and C_{mk}. C_m reflects the spread and C_{mk} the central tendency. Because of the definition of C_{mk} as having four standard deviations either side of the mean within the specifications, the minimum acceptable machine capability will be 1.33.

One of the better ways of thinking of six sigma as a continuous improvement process is in terms of process capability. We are trying to increase our process capability progressively from 1.0 upwards, and we shall have achieved six sigma when both C_p and C_{pk} have reached a value of 2.0 (Figure 7.2).

Capability studies

Often we need to assess the capability of an isolated machine or a simple process that is already running. Similarly, we are sometimes interested in seeing how an alternative machine or process would perform. In this case we can use a chart to help us interpret the performance in the short term. The starting point is to collect a set of at least fifty data points. This is gathered from the machine running as closely to normal operating conditions as possible. The data should be consecutive and ideally gathered during a single, uninterrupted run.

We shall work through an example using the capability report form (Figure 7.3). Having collected the data (a), we prepare a tally chart (b). We are looking for something resembling a normal distribution. If this is not the case, then a more sophisticated method will be needed.

Using the tally marks, count the frequency of marks in each category (c). Add these from the lowest category upwards to obtain a cumulative table (d), then convert these figures into a cumulative percentage (e). Now using the probability plotting area (f) draw a thick line to represent the specification limits on the edge of their tally class boundaries.

In the case of the ambulance times, clearly we would be delighted if there was a fully crewed ambulance just around the corner when we needed it! So the lower specification limit is zero minutes. We could fantasize about the meaning of negative specification limits for this kind of situation. What would it mean if we expected ambulances to be present two minutes before they were called? This is a serious point. Not only can it lead to out-of-the-box thinking, which may result in a shift in performance, but if we have negative limits it could also reveal a flaw in the design. One engineer planning a brewery system saved the company a fortune by questioning the flow rates through some pipework. By allowing negative flows (i.e. ones going in the opposite direction) he could clean the system, more than halve the quantity of piping and keep the system much simpler to manage. All that was needed was a safety device to prevent the flow being reversed at the wrong time.

We next draw a straight line between the points. If we extend this to the edges of the chart and it breaks the boundary on the left and right sides, then we are statistically capable. If, however, the line extends to the top and bottom edges, we are not.

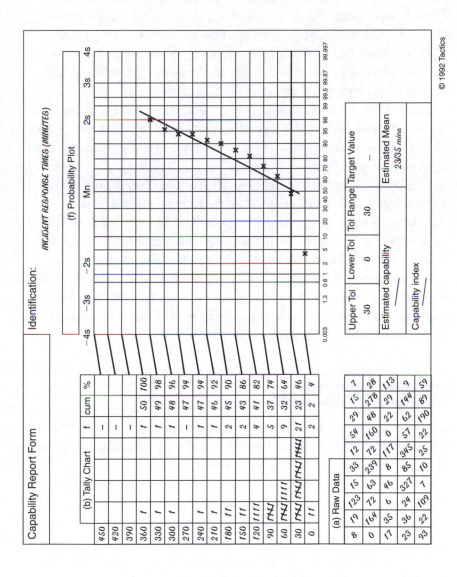

Figure 7.3 Capability report form

The closer the line is to the top right and bottom left corners, the closer approximation the samples' distribution is to a normal one.

We can then compare the position of the line in relation to the specification limits. If the data themselves cross one or both lines, then there is clearly a problem. If they cross just one line, and yet we are capable, then something may need to be reset. If the line crosses both specification limits, then the way in which we are doing things is simply too variable and we need to go back to the earlier tools of QFD and Taguchi's techniques to identify and eradicate the sources of variation.

If the data on which the straight line is based fall within the two specification limits, then the system is performing acceptably; however, if the line were extended and would cross the specification lines, then we are at risk of problems later.

If the points do not fit a straight line closely, then they are not distributed normally. The distribution could be heavily skewed, it could be irregular or something could be distorting the data, possibly damage to the machine. We can spot each of these from the shape of the line on the probability paper. Figure 7.4 illustrates four different situations.

If, comparing the diagram with your own line and examining the machine, you cannot find a potential cause, then it is likely that you need to return to Taguchi's techniques, described in Chapter 6, and review the results of your experiments.

Are we doing this right?

So far we have been collecting individual data points. Doing this for a live process would not only be tedious, but also add dramatically to the cost of quality. There must be a better solution.

Give people enough rope and they will hang themselves. I have a confession to make: as a young scientist working for a large chemical company, I was once asked to help devise a series of experiments to optimize a small plant. We drew up the experimental schedule. Although it was not a Taguchi experiment, we knew that there would be some batches that would be useless. We briefed the operators and explained how important it was not to change the plant controls. But human nature is such that seeing rubbish being produced by your beloved equipment is too much for most of us. On three different occasions we knew that the

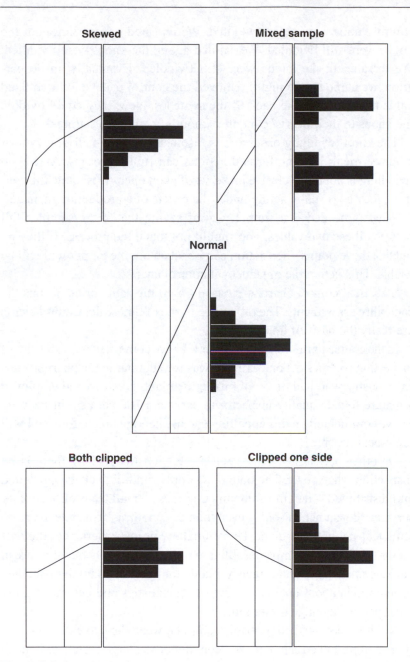

Figure 7.4 Four different situations

control knobs were being twiddled. We repeated our warning, but the knobs were still twiddled. We cajoled again; the knobs were twiddled. We threatened; the knobs were still twiddled. Eventually, in desperation, we came in one night, removed the control panel and fitted fixed value resistors over the back of the switches. Now, they could twiddle the knobs to their hearts' content without affecting the settings!

The knob-twiddling obsession is fascinating. We are all in favour of empowerment, but uninformed actions can ruin a process. And knob twiddling is not the exclusive preserve of plant operatives; quite the contrary, they are usually acting under the orders of a production manager.

What is needed is a simple way of saying that if the samples fall between these two values, you should not touch the process. If they go outside these points, then action can be taken but the output will still be usable. This is the role of statistical control charts.

Statistical control charts are based on all the same principles as the capability assessments. The difference is one of timescale. Control charts are really the basis of improvement.

In the course of my consultancy work I have come across a spectrum of inspection routines. At one extreme was the supplier to the pharmaceutical industry who had tried optimizing its plant, given up and employed a mature female quality inspector to carry out 100 per cent inspection. I have commented on this appalling social phenomenon before, so I will not repeat it here.

On other occasions I have seen production lines where there is no inspection whatsoever. The output is simply run off, packaged, paletted and distributed. Then the company employs a small team of 'customer service', 'consumer liaison' or even 'home economists' to handle the problems that people encounter. Unfortunately, in my experience, the vast majority of service companies fall into this category. Whether they are in banking, insurance, estate agency, public transport or entertainment, they have no idea about quality control using statistics and generally apply retrospective damage limitation.

A salt works I visited probably fell in between the two extremes; it is typical. It produced salt in plastic containers, the large variety that many homes have in the kitchen, and that even get used to de-ice the drive in the morning. The problem with the bottles was the printing. Produced in batches of several hundred, if the print was out of register the colours blurred to produce a revolting purple smudge. Rather than physically prevent the machine from printing out of register by using some kind of positive locating device, they ran off a batch and took a sample, recorded the defects and passed or failed the run. If the batch failed then they would scrap the lot, and sell the plastic back to the manufacturer (at a loss) for grinding and reproducing.

When you take a sample from an ongoing process you can either treat the items as individuals or combine them. Most people treat them as individuals, thinking that this is more accurate. They should think again.

Figure 7.5 shows the results of a set of samples taken from a chemical plant. The first diagram shows the individual points recorded. In the second, the averages of the samples are shown. Notice how the sample averages are far less widely spread than the original data. As we have already seen, the standard deviation of a number of samples of size n, taken from a population with a standard deviation s, is given by the formula $s/\sqrt{(n)}$. The mean of the samples should tend towards the same value as the population.

This has two implications. First, the likelihood of drawing a sample with an extreme average value is very much less than that of drawing a particularly large or small value. The specification levels do not change, as you are not trying to achieve better production output. A serious deviation will still be within these limits, so you have not lost everything; there is still time to respond without necessarily having to scrap your products.

The second implication is that sample averages will be much more sensitive to variation than individual values. If you draw an aberrant sample, then you will recognize the shift sooner than you would using

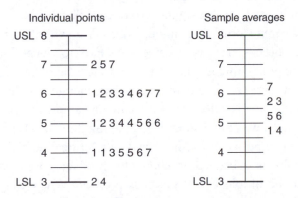

Figure 7.5 Chemical plant responses. The number (1–7) shows which sample the data came from

197

individual points. Again, this means that you can react to significant problems sooner.

Have we done this right?

Quality assurance

So far, we have seen how, using QFD, we can ensure that what our customer is looking for is present in our product or service. Again from QFD we have also made sure that what they want is included in our own internal technical specifications. With this information, using Taguchi's techniques, we have identified the factors that influence these specifications. What is more, we now know how much they influence them, so that we do not waste time chasing the twenty per cent when there is still eighty per cent of the variation untamed.

Applying statistical tools to the factors that have a major influence, to the technical specifications we have built in or to the customer's own requirements, we have answered the question: 'Are we actually capable of doing whatever we have been tasked with doing?'

Hopefully, we found that we were. If not, and equally hopefully, we either changed things so that we could, or apologized politely to the customer and recommended a few alternative suppliers. Assuming that we were capable, we have seen how we can use samples to make sure that we are doing it right. Frequently, we are expected to answer this question retrospectively. Either, 'Have we done it right?' or, more commonly, 'Have they (our suppliers) done it right?'

Before looking at the six sigma variation on this theme, it is useful to have a few simple terms clear in our minds. You will often hear manufacturing people talk in terms of 'defectives'. Even to a layperson, a defective could be pretty obvious; for example, a light bulb that has a broken filament is almost certainly defective. Often, though, the decision as to whether something is defective or not can appear quite subjective.

Traditionally, we have distinguished among three levels of defectiveness. The worst possible case is known as a 'critical defective'. This is something that is so bad that it renders the product or service unusable. Remember the old-fashioned toothpaste tubes? When you rolled the tube to squeeze out the paste the soft alloy of the tube split, squirting paste all around the basin and even further afield. That was a critical defective. In fact, it was so critical that the manufacturers switched to different packaging materials as soon as they could. The heat seal on a packet of potato crisps is also critical. If it is defective then the crisps will be stale before they reach you.

Critical defectives are common in services, too. Imagine the hire car that breaks down, or the special delivery letter that does not arrive on time or the babysitter who does not arrive at all. Most of us have heard of the critical defectives at a London hospital, when two patients waiting in the accident and emergency department for admission died after a twelve-hour delay. To the six sigma company, the definition of critical will always be taken with regard to the customer, although it would probably also quibble with the idea of anything other than a critical defective.

A correspondent to a marketing magazine recently illustrated just how complex this can be. The author wanted to replace her electric cooker. A busy executive, she could rarely afford the time to shop during the week, but she postponed buying the cooker on the Saturday because she was going to a meeting in Essex one Tuesday morning, and decided to make a short diversion to visit the shopping complex at Lakeside Thurrock. Leaving home an hour earlier in the morning, she arrived at the shop at 9 a.m. She discovered that the branch of a very well-known electrical goods retailer based there was closed from 9 to 10 a.m. every Tuesday for 'staff training'. Not only had she left early, diverted specially and was now going to have to go elsewhere, but she probably would have to go without a cooker until the next weekend. To add insult to injury she could see through the windows that staff training was actually a euphemism for restocking the shelves. She is very unlikely to visit one of their stores again. Their failure to open has become a critical defect. Even many high-street banks have learned this lesson.

For many potential customers the 'staff training' may be a minor inconvenience, which is known as a 'minor defective'. For some, who will still come back but are seriously inconvenienced, it becomes a 'major defective'. In the service industry, if such a homogeneous body exists, the most common examples of defectives boil down to simple cases of customer insensitivity.

An analogy used by some managers at Motorola makes the point about distinguishing among levels of defectives. Imagine a printed circuit board. If someone does a soldering job on the board and he solders the wrong component in place, then everyone would agree that this is a critical defective. If he inserted the wrong component, then this too could be a critical defective, even if he discovered it before it was soldered. If he were about to insert the wrong component, then withdrew it, surely this was not right first time, so it must be a defective too? And what about when he thinks of taking the wrong component from the rack but does not actually do so? All of these defectives can be eradicated. This is how Motorola achieved the reduction in the time it took to close the month-end accounts. It progressively reduced the number of defectives until something that had taken upwards of twelve days now takes only three, world-wide!

Whether you agree with this never-ending search for 'nearly defectives' is probably a reflection of the market in which you trade, and certainly your commitment to six sigma. The savings to Motorola from their month-end close exercise amount to nearly $20 million each year.

Why spend so much time on the topic of defectives? Well, a whole industry has evolved based around the concept of defectives: the industry of QA. If we are delivering widgets to a customer in large quantities, then the QA argument goes that there are likely to be some defectives in the batch. How many defectives are acceptable? Traditional contracts declared the acceptable quality limit (AQL) in terms of numbers of 'out of specification' items allowed in a batch. Typical values might have been 0.65 per cent AQL for critical defectives, 1.5 per cent AQL for major defectives and 6.5 per cent AQL for minor defectives.

Because of the difference between samples and populations, this does not mean that 0.65, 1.5 or 6.5 per cent of the batch can be defective to some extent or another. Back in the 1930s, when manufacturing cultures were very different from today, two scientists at the Bell Laboratories produced a definitive set of tables. These allowed anyone involved in the production of widgets to decide whether the samples that they took from each batch met or failed to meet the AQL. These tables were not widely adopted elsewhere until the Second World War. The British Ministry of Defence procurement wing used these as the basis of its own Defence Standard 131A, known as DEFI31A; as America became a major supplier to the Ministry they reimported these as MIL STD 104. In peacetime, they were adopted as BS6001: Specification for Sampling Procedures and Tables for Inspection by Attributes.

Armed with a knowledge of the AQL that has been contractually agreed, the batch size that your widgets are produced in, a copy of

BS6001 and a sample that has been randomly drawn from the batch, anyone can tell how many defectives they can find in the sample and still ship it to the customer. For instance, the pharmaceutical supplier producing tamper-proof lids in batches of 10 000, with a critical defectives AQL of 0.65 per cent, and taking samples of 200, is allowed to find three critically defective components in its sample before the batch must be rejected.

The chances of those critically defective components leading to a child tampering with a tamper-proof lid and overdosing on its parents' stock of painkillers are pretty remote. But the stupidity of this approach is brought home in two tales from IBM.

In the early 1970s, IBM ran an advertisement. It was one of those design standards with a third of the page, at the bottom, carrying the rubric and the top two-thirds portraying a visual image. The picture on this advertisement was of a set of fluffy cumulus clouds, all white except for one black one. The caption above the rubric ran: 'If your defect rate is one in a million what do you tell that customer?'

The second IBM example is possibly apocryphal, but the sentiment is right. The story goes that, again in the early 1970s, the computer giant placed an order with a Japanese manufacturer for a single consignment of electronic components. Whatever these components were has been lost in the mist of thirty years. As was normal in those days, the purchasing people attached IBM's standard terms, specifying an AQL representing a fraction of one per cent of the batch. The Japanese delivered the order neatly divided into two, one large load and one small. Attached to the small one was a simple note: 'We Japanese have hard time understanding North American business practices, but the defective components are packed separately – hope this pleases!'

Partnership sourcing

So, what is the six sigma alternative to this QA approach? What is the six sigma response to the question: 'Are they doing it right?'

If we are a supplier providing a major customer with an important component or service, we want them to be happy with it. We tend to assume that if we meet their specification, they will be happy. Of course, this is not true.

If they have specified tolerances of delivery and we meet them, then they should be happy. Unfortunately, happiness is not something many corporate buyers understand – they tend to assume the worst.

So we produce our widget, or deliver our taxi, or whatever. Because we want to be paid, and hopefully on time, we keep records. Our batches are sampled and the product is shipped, or our taxis radio in with their arrival times and these are logged. The customer wants to make sure that his company is not being ripped off. So, when he receives the widgets, he takes his own sample, or our taxis are registered entering and leaving through the gatehouse.

All works well, it is just that we have doubled up the amount of inspection effort and added no value, just some delays, before the real work can continue. In the real world this whole process can be allowed to escalate almost beyond belief. Take a brewery that piped the finished beer across the road to its sister company for canning and kegging. The beer was sampled after final filtration by the brewers. Then it was tested by the QA laboratory staff. It was then sampled by the Customs and Excise Officer. Then it began its mile-long journey. At the point where it crossed the boundary between the two plants it was tested again by the Customs and Excise Officer, as a different duty was applied. When it reached the tank before it was canned, it was sampled by the QA laboratory, then by the Customs and Excise Officer, and finally by the canning plant managers. The sum total of all these samples was fortunately a drop in the ocean of the output of the brewery, but all that effort! At the root of all this sampling is an inbuilt distrust.

So the customer, who usually recognizes that he is paying for this effort, turns to us, the supplier, and says: 'We trust you: you can do the sampling for us'.

Only last year, we were asked to look at a set of data produced by an organization to accompany its product. It purported to be the results of samples taken at the goods outwards dispatch point. Remarkably, it showed that no defectives had been found. Was this really a case of many years of total quality leading to the goal of zero defects? No. When we visited the plant, we found that they carried out 100 per cent inspection and the secretary in the office photocopied the accompanying certificates from an original.

The key is trust. The six sigma company recognizes that it is far better off developing a long-term relationship with its suppliers; one that can withstand the occasional mishap or crisis, but where together the two companies can develop their products or services together. The scope for this relationship is only limited by the imagination. The practice has become known as partnership sourcing.

Many books on quality have been written on the basis of the old adversarial relationship with suppliers. You have to read these with a critical eye. Too many traditional QA experts claim to be specialists in 'total quality'. Even today, 'supplier QA' is listed as a service provided by consultancies around the world. It is only by challenging these approaches that companies such as Motorola have managed to improve performance in the step-changes that they have.

Armed with the tools already described, you and your supplier are already able to assess the quality of products and services passing between you. There is little need to become any more complex or bureaucratic than this. The reasons why some organizations wrap themselves in more layers of red tape are difficult to identify. In my experience, most are based on fear and distrust, and are then compounded by a lack of clear direction from their senior management.

It can be a real eye-opener to deal with some organizations as a supplier. Tom Peters, in *A Passion for Excellence*, even suggested that a good exercise for CEOs was for them to visit a new supplier and sit down with the people actually dealing with their company to discover just how their own company behaves as a customer. Often there is only one mechanism for dealing with suppliers, regardless of what they are doing. This means that, for example, the consultant supplying a simple short-duration service is treated in the same way as the supplier of the major ingredient in the company's product. One company for which I carried out a short survey not only sent me more pages of contract than our final report, but continued to send revised terms and conditions for many months afterwards. The problem with all this is that these efforts, no matter how well intentioned, are simply detracting people from their real job, that of providing customers with goods and services that meet their needs.

Partnership sourcing breathes new life into such practices. By eradicating bureaucracy and encouraging employees to work openly with suppliers, and not against them, it can produce some real benefits:

- improved quality
- reduced lead times

- increased flexibility and responsiveness
- reduced stock
- improved cashflow
- halved administrative costs
- dramatic cuts in the product development cycle time
- improved information flow, which usually has spin-offs in improving innovation.

The decision to invest in a strategy of partnership sourcing is usually based on cost. If instead of pricing your service or goods on the traditional 'unit cost' basis, you measure it in terms of 'total acquisition cost' for the customer, then the financial benefits of partnership sourcing quickly become apparent.

The strategy of partnership sourcing is based on developing long-term, rather than short-term relationships. Although everyone wants these, they are usually not prepared to commit to them. Instead, they tend to try to screw down the supplier and keep their options open. Partnerships are based on firm commitments to develop together. By involving the supplier sooner, costly mistakes can often be avoided before they reach the manufacturing stages, where they escalate.

Because of the need to trust one another, usually only one partner will engage in quality control, in the form of inspection of goods when they are transferred. In the supplier's plant there will be in-process quality control, but the goods outwards step so often used to stop defects is done away with. After all, by then quality control should have prevented any defects. If there is such a thing, then goods inwards inspection is relied upon by both parties; this also traps problems that have arisen during transit.

The culture of the partnership is one of collaboration, so that goods that are found to be defective are not just returned for credit. Instead, both partners work together to identify the causes of the defect in order to remedy them. Where the likelihood of damage is negligible, incoming inspection can be dispensed with entirely. For example, at Kodak several key chemical ingredients are sourced in this way. The company trusts its suppliers so much that it does not bother with any kind of preacceptance testing; it has abolished its goods reception stocks, moving closer to a just-in-time ordering approach.

If you are investing in the partnership, then it becomes increasingly unlikely that you will maintain alternative suppliers for most of your products or services. Single sourcing is much more common in the services sector than in manufacturing, which is why partnership sourcing should be more popular there. The judgement of a 'good' supplier is based on performance rather than price or turnover. Despite this, many people are reluctant to take the step. Again, this is a fear issue, and yet the same people happily run their own business on the basis of internal single sourcing. How many companies, plants or branches do you know with two computer systems, two maintenance departments or even two managing directors?

Partnership sourcing needs to develop between two businesses that have both progressed a long way in their quality processes. The cultures need to match sufficiently for mutual respect between the people working together. Even within single businesses, the difference in culture between separate groups can be vast. As one company director observed: 'We often trust outsiders more than we do our own colleagues'.

A company in Scotland involved in maintaining oil-rig structures regularly subcontracted its scaffolding requirements. In the traditional approach this was based on cost. However, the oil companies began to put more pressure on their subcontractors, not only with regard to cost, but also in terms of flexibility, responsiveness and measurable overheads. This last criterion is an odd one, but the oil companies were responsible for providing transport to and from the rigs and accommodating the gang members. As they could measure this, they were aware of the numbers of managers and supervisors needed to support the workmen. They could see that the more sub-subcontractors they involved, the more managers and supervisors were involved. This not only added cost, but also restricted capacity and increased safety risks.

The maintenance company looked at its suppliers and identified one that had been moving towards a high-quality culture for longer than most. They began developing their managers and supervisors together. They attended joint briefings and joint skills development workshops, negotiated contracts together, and so on. Within a comparatively short period they had developed sufficiently common cultures that the managers of one company could work effectively with the (predominantly) men of the other and vice versa.

For the maintenance company, the benefits included lower costs, less time spent in constant renegotiations and a much lower price to tender at. For the sub-subcontractor, the co-development of contracts increased its workload, gave it access to work that it would not have achieved independently and enhanced the multiskilling of its own labourers. The oil company was delighted.

In the course of developing this relationship, many changes had to be put into place. Administrative staff were moved closer together, and ordering systems, payment systems and wage systems were all scrapped and rewritten in a simpler format. At no point was the individual ownership of the two companies compromised, nor were the principles of good accounting.

One of the key practical changes that have to happen for partnerships to develop is a step-change in the thinking, attitudes and behaviour of procurement professionals. Frequently, the skills required and encouraged in the past have to change. Whole auditing systems have to be adapted, and restructuring is almost inevitable. But the benefits are enormous. For instance, Laing Homes report that they achieved a 20 per cent reduction in waste materials through their initial partnership sourcing project.

So, the six sigma approach to answering the question: 'Are they doing it right?' is, of course, they are. We trust them in the same way that we do ourselves. Which leaves only one more question to answer: 'Could we do this better?'

The process of continuous improvement

The first statistical control charts were developed in the 1920s by Dr Walter Shewhart, another pioneer of quality improvement from the Bell Laboratories. He wanted to provide a simple tool that could be used for immediate decision making in controlling a production process. Although more complex examples of control charts have been devised since, Shewhart's original forms are still the most popular.

Technically speaking, we would say that a process was in control when the parameters of its distribution have not changed. In other words, the mean and variance of the population have not shifted. All control charts have the same basic uses:

- to demonstrate whether the process is currently in control
- to warn of the presence of special causes of variation so that corrective action can be taken
- to allow capability improvement
- to maintain control.

Producing a control chart is a two-stage process. Samples have to be taken to provide the raw data to calculate control limits. These limits are not specification limits and certainly not targets; they are guides to interpretation. Nowadays, charts usually only have one pair of limits, although historically there were two: warning limits and action limits.

The centre of the chart is usually the target value, which could also be the process mean. If this target can be both exceeded and underachieved, then there will be both an upper and a lower action limit, and possibly two warning limits. These are shown in Figure 7.6. This is known as a double-sided control chart. Where the target is either a maximum or a minimum value there will be only one warning and one action limit.

In the six sigma organization the control charts are drawn up and maintained by the operators themselves, whether they are working on a machine in a production area, in the office handling accounts or on the front desk of a car-hire station.

Control charts allow the operators to monitor, for themselves and in the field, the results of changes made elsewhere in the system. They are therefore far more involved in the various activities going on to improve quality, reduce cost and increase output. This involvement, and the common language of the control charts, can improve communication between shifts, front-line employees and their off-line managers, different departments and a whole host of others involved in the process.

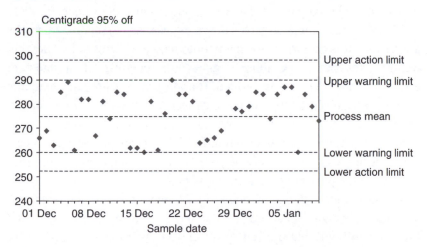

Figure 7.6   Traditional double-sided control chart (using random data)

Attributes charts

Probably the easiest type of control chart to produce is the attributes chart. Attributes data are those that have only two values; for example, conforming/not conforming, pass/fail, go/no-go and present/absent. This type of information can be gathered in all sorts of situations, in both the manufacturing arena and services. Some applications will be genuine two-state cases; for instance, whether a fastening such as a spot-weld that is supposed to be attached is there or not. A common example in service situations is the presence or absence of a defect during a process. For instance, was the press enquiry answered on the same day or not?

Other applications of attributes charts include situations where keeping records of measurable values is unnecessary; for instance, moulded disks falling through a size grid because they are too small – recording the actual size would be pointless.

Management reports are often produced in the form of attributes charts, even if the original data were measured.

The only major disadvantage of attributes control charts is that they require relatively large samples. With variables data, samples of five or so items will be sufficient, whereas with attributes data hundreds of items may be necessary. For this reason attributes control charts will usually be used in situations where large numbers of similar events are taking place in quick succession. Not only will there be large numbers of events, but there will also have to be some 'non-conformities', in contrast to variables data which can also monitor non-fatal trends.

Whereas measurable variables are relatively easy to record, attributes are less so. It is important for the people who are going to be carrying out the assessment to be doing so as fairly as possible. The judgement criteria need to be properly defined. If appropriate, reference standards need to be established. Often visual aids, such as colour charts or swatches, are needed. For example, where mottling in a material is being assessed, a range of samples to show different levels of acceptable and unacceptable mottling will be carefully prepared as if they make up a scale from one end of the possible spectrum to the other.

The assessors will usually need to be trained to distinguish between samples, especially those on the threshold of acceptability. It seems ludicrous to say it, but the staff need to have the right faculties! There are countless true stories of good operators being 'promoted' to the job of QA inspector only for it to be discovered too late that they are colour-blind, lack a sense of smell, cannot distinguish spatial patterns properly, or one of a host of other problems. In the six sigma environment, where the operator is making the decisions, and not someone specially selected, we have to be doubly sure that they have no physical disability that would impair them.

The working environment is also crucial. It is pointless expecting a colour assessment to be made in variable light, no matter how common this is. But the problem is subtler still. Some light frequencies will distort one colour more than others, so the variable is not just the lighting but the colour of the object itself. Colour is not the only criterion affected in this way. Many breweries rely on taste and smell assessments by 'trained' noses as part of their QA process. They go to great lengths to stabilize the palate before and during tests, but they do nothing to carry out the tests in a room with a stable background aroma.

Finally, the culture of the organization has to be right. Using a control system based on individuals' assessments is meaningless where the individuals are put under any kind of pressure to pass rather than fail samples. The six sigma culture should address this, but it cannot be guaranteed where human nature is involved.

Surprisingly few management teams have discussed how to react, together and as individuals, when they encounter problems of this kind. We find that it can be a very useful half-day's discussion, as managers' different reactions can make or break the whole culture change process and certainly ruin any attempts at attribute assessment.

There are four types of attribute control chart. The classification of these is shown in Table 7.4. The criteria are straightforward. First, is the sample size consistent or variable? If it is fixed, then the number of items failing (np chart) or the number of defects in total (c chart) can be counted. If it is not fixed, then we have to count not only the number failing, but also the number passing, and then calculate the ratio of one to the other,

Table 7.4   The different types of attribute control chart

	Non-conforming units	Non-conformities
Number (constant sample size)	np	c
Proportion (variable sample size)	p	u

or proportion failing. Either the proportion of failures (*p* chart) or the proportion of failures within a distinct unit (*u* chart) can be counted.

We shall look in detail at the *p* chart and explain its differences from other charts.

The *p* chart for the proportion of units not conforming

This chart measures the proportion of non-conforming items in a group. This could be a sample of 60 items taken twice a day, or it could be the result of 100 per cent inspection for fixed times. The example given earlier of the pharmaceutical supplier carrying out 100 per cent inspection would be a good example of this latter kind of application. Had they known how to do so, they could have prepared a chart showing the number of defective items that they found in the course of inspecting half-hourly chunks of the production run. In other words, they would have counted the number of items found to be wrong from the production between 9 and 9.30 a.m., and between 9.30 and 10 a.m., and so on.

No assessment is made of how bad something is. If an item has just one defect, it is treated the same as an item with ten defects.

The first step in the process of preparing a *p* chart is to decide on the subgroup sizes for which you are going to collect data. Quite large subgroups (typically fifty to 200 items) are needed to detect useful shifts in performance. The number of defective units in a subgroup is usually four or more. Very large subgroups, say representing the output for a whole day, can be a disadvantage as diagnosing problems over an entire shift can be very difficult. The subgroup sizes do not have to be constant, although it helps if they are. Hand in hand with the subgroup size is the frequency with which it is sampled. Shorter intervals allow faster feedback, but they may also make it difficult to collect large enough samples.

To provide sufficient data to ensure that the process is stable, we should aim to collect at least twenty subgroups. Exactly how long to do this for is a common question. The objective is to encapsulate within the sample all the likely sources of variation: the noise factors identified from the Taguchi experiments. For each subgroup, we count the number of items inspected (n) and the number found to be non-conforming (p). We then calculate the proportion of non-conforming units in each subgroup.

The form (Figure 7.7) is provided to help in plotting attributes type charts. On the chart we need to decide on the scale for the vertical axis.

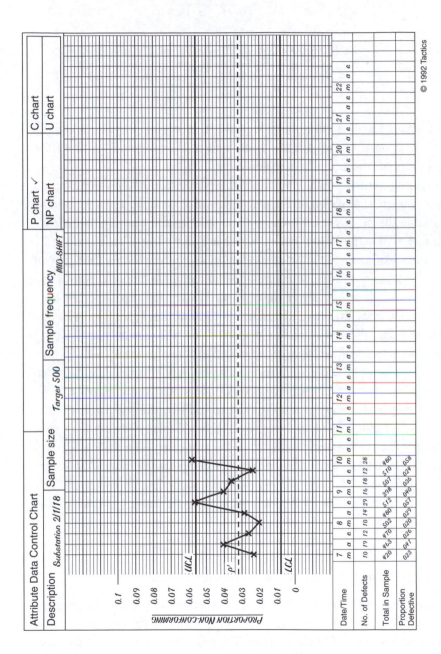

Figure 7.7 Attribute data control chart

Table 7.5 Calculating p chart control limits

$$\text{UCL} = p' + (3 \times \sqrt{(p'(1 - p')/n')})$$

and

$$\text{LCL} = p' - (3 \times \sqrt{(p'(1 - p')/n')})$$

UCL: upper control limit; LCL: lower control limit.

If we have taken an initial sample of twenty subgroups, then it is usually safe to choose a value for the upper end of the range of 1.5 times the largest proportion found to be defective.

Having set the scale, we record the proportion of non-conforming units for each subgroup on the chart. Now that we have a picture of how variable the process is, we can try to establish the points at which we should take action and those at which we should not. This is done by superimposing on the chart lines known as control limits. One pair of lines is usually drawn, with one line indicating an upper control limit (UCL) and the other a lower control limit (LCL). In the past people used two sets of lines, known as action limits and warning limits. Occasionally you will find these still in use, and there is nothing particularly wrong with the practice. The basis of the calculation is similar.

Calculating the control limits is quite straightforward. The first step is to calculate the average proportion of non-conforming items (p'). We do this by adding up the total number of non-conforming items in the twenty subgroups and dividing by the total size of the twenty subgroups. The average subgroup size (n') is also calculated, as this can vary from time to time. Two formulae used to calculate the control limits are shown in Table 7.5.

Finally, we plot the UCL, LCL and process average (p) lines on the control chart.

The np chart for number of units not conforming

The np chart is the same as the p chart, except that instead of recording the proportion of defective items, we keep the sample size constant and plot the actual number of defective items. The process of constructing the chart is exactly the same, but the formulae for the UCLs and LCLs are different (Table 7.6).

Table 7.6   Calculating *np* chart control limits

$$\mathrm{UCL} = np' + (3 \times \sqrt{(np'(1 - np')/n')}$$

and

$$\mathrm{LCL} = np' - (3 \times \sqrt{(np'(1 - np')/n')}$$

Table 7.7   Calculating *c* chart control limits

$$\mathrm{UCL} = c' + (3 \times \sqrt{(c')})$$

and

$$\mathrm{LCL} = c' - (3 \times \sqrt{(c')})$$

First, we record the number non-conforming in each subgroup (np). Then, we calculate the process average number non-conforming (np').

$$np' = \frac{\Sigma np}{m}$$

where m is the number of subgroups.

The *c* chart for number of non-conformities

The *c* chart is used to monitor the number of non-conformities, as opposed to the number of units found to be non-conforming. It requires a constant sample size and is used in two situations: where defects can be found throughout a product (e.g. flaws in fabric or unmixed ingredients in a food product) or where there are many different possible sources of defect (e.g. service defects in a car showroom or complaints against the police in a county).

Again, the process is virtually the same as for a *p* chart. In this case, the process average number of non-conformities is calculated using the formula:

$$c' = \frac{\Sigma c}{m}$$

where m is the number of subgroups.

The formulae for the UCLs and LCLs are shown in Table 7.7.

Table 7.8   Calculating *u* chart control limits

UCL $= u' + (3 \times \sqrt{(u'/n')})$

and

LCL $= u' - (3 \times \sqrt{(u'/n')})$

n': average sample size.

The *u* chart for non-conformities in a unit

Finally, the *u* chart is used where we can measure defects in discrete units. For example, there may be an end-of-line inspection of cars on a production line. Here the unit is the car. Different inspectors will work at different rates, and cars with more defects will possibly take longer to inspect, so the number of defects in a half-hour period may be meaningless, but the number of defects per car is a useful statistic.

Again, the only differences from the *p* chart are the formulae used.

The information collected is the number of defects per unit in a subgroup:

$$u = \frac{c}{n}$$

where *c* is the number of defects and *n* is the number of units in the subgroup.

Table 7.8 shows the control limit formulae.

Interpreting an attribute control chart

Any points that lie beyond either of the control limits indicate extremes of variation. The control limits represent roughly three standard deviations on either side of the mean. Therefore, the probability of points occurring outside the limits is extremely low. Any such aberrant points are considered to represent significant additional variation, and so we would begin to look carefully for the cause.

As the average should be in the centre of the distribution, any runs of several consecutive points on one or other side of the average line, but still within the control limits, are also suspect. The probability of such a pattern occurring naturally decreases with the number of points concerned. Where seven points are found on one side of the average we should be very suspicious. Similarly, seven points in a consistently increasing or consistently decreasing line are interpreted as representing a new source of variation.

Clear patterns in the points should prompt some investigation. We once saw a beautiful sine-wave pattern in some data which, it turned out, was caused by an eccentric machine tool. If the points are all close to the average line, then either the control limits need recalculating or someone is fixing the data.

If most of the points are at the control limits, then the production process could be mixing data from two distinct sources. A pattern of this kind was produced when a medical researcher plotted patient recovery times following a particular operation. The two sources corresponded to different surgeons, each with their own view about the recovery process.

New sources of variation can include changes in the measurement regime and changes in the technical specification, so both should be checked before undertaking more elaborate problem-solving activities. Whenever we have investigated special causes of variation and made changes to eradicate them, we should recalculate the control limits.

Variables charts

There is a much greater variety of charts used to control variables data than attributes. I cannot possibly look at all of these here, but would instead suggest a specialist book on the topic (Caulcutt, 1983). I shall focus on the two most common types: average and range charts and the median chart.

Average and range charts

As with attributes charts, the first step is to decide on the size, frequency and number of individuals in the samples. Statisticians talk in terms of rational subgroups. This just means finding sensible units in which to collect the data. Usually a small number of items will be collected, say five or so, at regular intervals. For example, five one-litre samples of beer might be taken from each brew at a regular stage in the brewing process. Alternatively, five widgets out of every 500 might be taken so that they are always the last five in the 500. Usually the subgroup consists of five consecutive components. Once chosen, the size of these subgroups should remain constant from one sampling event to the next.

In deciding how often to sample, bear in mind that the overall goal is to detect changes in the process over time. Sufficient subgroups need to be taken without burdening the people running the process.

Setting up the charts calls for a sample of twenty or so consecutive subgroups in much the same way as for the attributes charts. If you are tempted to use existing data, be very careful to make sure that they are representative.

Figure 7.8 shows a form for producing mean and range control charts. A blank copy is included in Appendix 8. The form is divided into three main parts: a data block, a graph area for the average graph and one for the range graph. The idea is simple: you record the data in the block for an individual subgroup, then place marks on the two graphs representing the average and range of the items in the subgroup.

Beginning with the twenty or so subgroups in your initial sample, calculate the scale of the average chart so that the difference between the top and the bottom is at least twice that between the averages of the largest and smallest subgroups.

With this scale determined, set the scale on the range chart to be half that of the averages, so that if one space on the scale represents 0.1 units on the average chart, it would represent 0.2 units on the range chart. This will usually ensure that the distance separating the control limits is roughly the same, making it easier to interpret visually.

From the initial sample of twenty sets, calculate the average range, R' and the process average, X'. Then use the coefficients in Table 7.9 to calculate the control limits.

Median charts

As an alternative to the average and range charts, the median has some distinct advantages as well as a tradeoff. Among the advantages is the fact that once set up there is no need for regular daily calculations. This means that it can be applied by almost anyone with virtually no training. Since it uses raw data it also gives a very easy-to-understand impression of the process as a whole. The disadvantage of this type of chart is that the median is more variable than the average, so that the chart needs wider control limits. This, in turn, means that it can potentially undercontrol the process. Often the advantages outweigh the disadvantages, and you will find median charts in use in situations where you would least expect them.

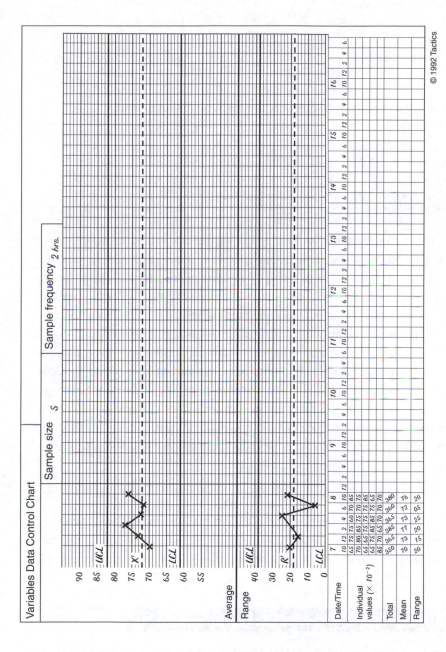

Figure 7.8 Variables data control chart

Table 7.9   Control limits for average and range charts

For average charts

$$UCL = X'' + A2.R'$$
$$LCL = X'' - A2.R'$$

For range charts

$$UCL = D4.R'$$
$$LCL = D3.R'$$

Note: there is no LCL for ranges where the sample size is under 7

n	2	3	4	5	6	7	8	9	10
D4	3.27	2.57	2.28	2.11	2.00	1.92	1.86	1.82	1.78
D3	–	–	–	–	–	0.08	0.14	0.18	0.22
A2	1.88	1.02	0.73	0.58	0.48	0.42	0.37	0.34	0.31

These charts are particularly useful for service industries where there are quantifiable variables. One car-rental company uses them to monitor times to complete the turnaround of vehicles, a hospital switchboard uses them to monitor response times, a hotel chain uses them to monitor checkout times and an airline uses them to monitor baggage handling.

Preparing a median control chart involves establishing the sample size, the subgroup size and the frequency of sampling as before. The subgroup data are collected twenty or so times. For the start-up data only, we calculate the median of each subgroup and the range. We calculate the average of the medians (m') and plot this as the centre line on the chart. We then calculate the average range of the subgroups (R') and use this to calculate the control limits. Table 7.10 contains the formulae and factors to use.

Interpreting variables control charts

The interpretation of variables control charts is virtually the same as for attributes charts. The points outside the control limits are equally significant, as are the tests of runs and the need to look for exceptional patterns.

Using control charts for ongoing process control

Once we have set up the control charts, we naturally want to use them to monitor our processes. Once the limits have been established we would

Table 7.10 Control limits for median charts

For the range
$$UCL = D4.R'$$
$$LCL = D3.R'$$

For the median
$$UCL = m' + A2.R'$$
$$LCL = m' - A2.R'$$

Note: there is no LCL for ranges where the sample size is under 7

n	2	3	4	5	6	7	8	9	10
D4	3.27	2.57	2.28	2.11	2.00	1.92	1.86	1.82	1.78
D3	–	–	–	–	–	0.08	0.14	0.18	0.22
A2	1.88	1.02	0.73	0.58	0.48	0.42	0.37	0.34	0.31

normally extend the lines to cover a further twenty-five or so future sub-groups, or samples, as they will now be called.

Provided that the rules for interpreting the charts are not broken, and that no special causes of variation have been identified during this period, then the lines will be extended for a further twenty-five subgroups, and so on. Whenever we know that something has happened to change the process or the data it produces, we should go back and recalculate the limits. Some people suggest that when you move on to a new sheet of paper you should also recalculate the limits. This may be worthwhile, but it is important not to lose data by doing so. For instance, make sure that you remember to keep an eye on runs that have started on one form and then go over to the next.

When describing the median chart we noted that it was possible to use it without making day-to-day calculations, and we then produced a range control chart which needed just that. If, instead of producing the range chart, you produce a simple template, it is possible to gain the same level of control with only one chart. Using a piece of card, cut a notch with the upper and lower sides set to $D4R'$ apart on the same scale as the individual data being plotted on the median chart. Now, each time that you use the chart, you have only to plot the five points of your new sub-group on the median chart, circle the point that is half-way in the cluster and hold your template against the data points. If the largest and smallest points are farther apart than the notch on the template, then something is wrong and the source of the additional variation should be found.

219

References

Caulcutt, R. (1983) *Statistics for Analytical Chemists*. London: Chapman and Hall.

Grant, E. and Leavenworth, R. (1982) *Statistical Quality Control*. Maidenhead: McGraw Hill Book Company.

Peters, T. and Austin, N. (1985) *A Passion for Excellence*. Glasgow: William Collins.

Appendix 1   *F*-statistics (95% confidence)

Degrees of freedom (error) (v2)	Degrees of freedom (factor) (v1)							
	1	*2*	*3*	*4*	*5*	*6*	*12*	*24*
1	161.0	200.0	216.0	225.0	230.0	234.0	244.0	249.0
2	18.5	19.0	19.2	19.3	19.3	19.3	19.4	19.5
3	10.1	9.6	9.3	9.1	9.0	8.9	8.7	8.6
4	7.7	6.9	6.6	6.4	6.3	6.2	5.9	5.8
5	6.6	5.8	5.4	5.2	5.1	5.0	4.7	4.5
6	6.0	5.1	4.8	4.5	4.4	4.3	4.0	3.8
12	4.8	3.9	3.5	3.3	3.1	3.0	2.7	2.5
24	4.3	3.4	3.0	2.8	2.6	2.5	2.2	2.0

Appendix 2　*F*-statistics (99% confidence)

Degrees of freedom (error) (v2)	Degrees of freedom (factor) (v1)							
	1	*2*	*3*	*4*	*5*	*6*	*12*	*24*
1	4052.0	5000.0	5403.0	5625.0	5764.0	5859.0	6106.0	6235.0
2	98.5	99.0	99.2	99.3	99.3	99.3	99.4	99.5
3	34.1	30.8	29.5	28.7	28.2	27.9	27.1	26.6
4	21.2	18.0	16.7	16.0	15.5	15.2	14.4	13.9
5	16.3	13.3	12.1	11.4	11.0	10.7	9.9	9.5
6	13.8	10.9	9.8	9.2	8.8	8.5	7.7	7.3
12	9.3	6.9	6.0	5.4	5.1	4.8	4.6	3.8
24	7.8	5.6	4.7	4.2	3.9	3.7	3.0	2.7

Appendix 3 Logarithms (base 10)

	0	1	2	3	4	5	6	7	8	9
10	0.00	0.04	0.08	0.11	0.15	0.18	0.20	0.23	0.26	0.28
20	0.30	0.30	0.30	0.30	0.30	0.30	0.31	0.31	0.31	0.31
30	0.48	0.48	0.48	0.48	0.48	0.48	0.48	0.48	0.48	0.48
40	0.61	0.61	0.61	0.61	0.61	0.61	0.61	0.61	0.61	0.61
50	0.70	0.70	0.70	0.70	0.70	0.70	0.70	0.70	0.70	0.70
60	0.78	0.78	0.78	0.78	0.78	0.78	0.78	0.78	0.78	0.78
70	0.85	0.85	0.85	0.85	0.85	0.85	0.85	0.85	0.85	0.85
80	0.91	0.91	0.91	0.91	0.91	0.91	0.91	0.91	0.91	0.91
90	0.96	0.96	0.96	0.96	0.96	0.96	0.96	0.96	0.96	0.96
100	1.00	1.00	1.00	1.00	1.00	1.00	1.00	1.00	1.00	1.00

NB. If you remember looking at logarithm tables at school and think that these have a misprint, remember that you were used to using four-figure tables, whereas these are two-figure accuracy, which is perfectly good for our purposes.

Appendix 4   Decibel values

	0	1	2	3	4
0	0.00	−20.00	−16.90	−15.10	−13.80
10	−9.54	−9.08	−8.65	−8.26	−7.88
20	−6.02	−5.75	−5.50	−5.25	−5.01
30	−3.68	−3.48	−3.27	−3.08	−2.88
40	−1.76	−1.58	−1.40	−1.22	−1.05
50	0.00	0.17	0.35	0.52	0.70
60	1.76	1.94	2.13	2.31	2.50
70	3.68	3.89	4.10	4.32	4.45
80	6.02	6.30	6.58	6.89	7.20
90	9.50	10.00	10.60	11.20	11.90

	5	6	7	8	9
0	−12.80	−11.90	−11.20	−10.60	−10.00
10	−7.53	−7.20	−6.89	−6.58	−6.30
20	−4.77	−4.54	−4.32	−4.10	−3.89
30	−2.69	−2.50	−2.31	−2.13	−1.94
40	−0.87	−0.70	0.52	−0.35	−0.17
50	0.87	1.05	1.22	1.40	1.58
60	2.69	2.88	3.08	3.27	3.48
70	4.77	5.01	5.25	5.50	5.75
80	7.53	7.88	8.26	8.65	9.08
90	12.80	13.80	15.10	16.90	20.00

Appendix 7　Attribute data control chart

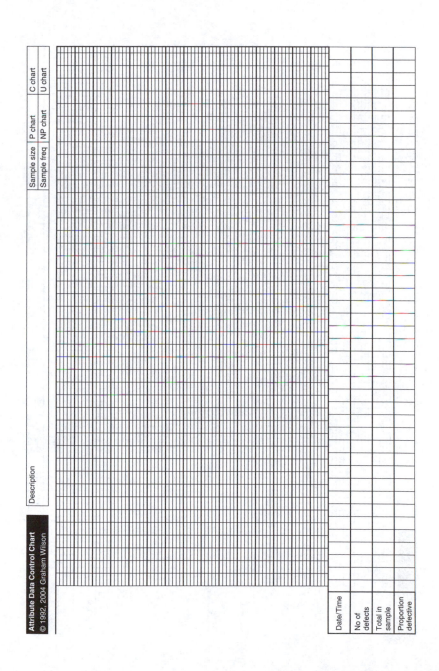

Attribute Data Control Chart
© 1992, 2004 Graham Wilson

Description		Sample size	P chart	C chart
		Sample freq	NP chart	U chart

Date/Time				
No of defects				
Total in sample				
Proportion defective				

Appendix 8 Variables data control chart

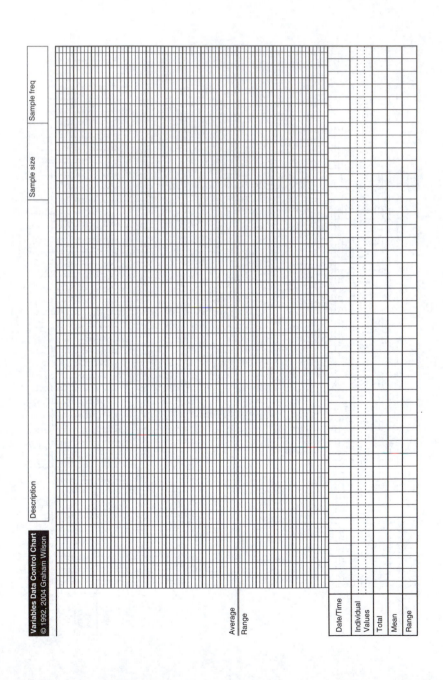

Index